2/96

D1008862

THE TOTAL PACKAGE

ALSO BY THOMAS HINE

Populuxe
Facing Tomorrow

THE TOTAL PACKAGE

The Evolution and
Secret Meanings of
Boxes, Bottles, Cans, and Tubes

THOMAS HINE

LITTLE, BROWN AND COMPANY
Boston   New York   Toronto   London

Copyright © 1995 by Thomas Hine

All rights reserved. No part of this book may be reproduced in any form or by any electronic or mechanical means, including information storage and retrieval systems, without permission in writing from the publisher, except by a reviewer who may quote brief passages in a review.

First Edition

All illustrations are from the author's collection with the exception of the following:
Rosalind Berman collection, photographed by Michael Mally: 2A, 9A, 12A, 12B, 12C, 15A, 15D, 15E
Campbell Soup Co.: 6C, 12D, 13E, 20B
Cheskin + Masten: 7B
Gerstman + Meyers: 5B, 24B
Irv Koons collection: 15B, 15C, 17D, 18A, 20A, 22C, 22D, 22E, 22F, 23H, 23I, 23J
Philadelphia Museum of Art: 22A
David Slovic: 8A

Library of Cataloging-in-Publication Data

Hine, Thomas
    The total package: the evolution and secret meanings of boxes, bottles,
cans, and tubes / by Thomas Hine.
        p.   cm.
    Includes bibliographical references.
    ISBN 0-316-36480-0
    1. Packaging — Social aspects — United States.   2. Packaging — United
States — Psychological aspects.   3. Advertising — Social aspects —
United States.   I. Title.
HF5770.H56   1995
658.5'64 — dc20   94-36445

10   9   8   7   6   5   4   3   2   1

MV—NY

Published simultaneously in Canada by Little, Brown & Company (Canada) Limited

PRINTED IN THE UNITED STATES OF AMERICA

658.564
HIN
1995

To James Chan
Accept No Imitations

Contents

Foreword

When you write a book about packaging, you can't simply rely on interviews and library research. You have to go to where all those boxes, bottles, cans, tubes, pouches, bags, and hanging plastic clamshells dwell. You have to go to the store.

In the interest of research, it was off to my local Super Fresh and Acme supermarkets, to the Rite-Aid pharmacy, Sam's Club, Computer City, Home Depot, and Tower Records and many other specialty shops and superstores.

I tried to remain detached. I took note, for example, of new, less wasteful, and more convenient ways of packaging drain cleaners, painkillers, and cranberry juice. I studied the colors and typography that define the new, upscale private grocery brands. At the warehouse club, I marveled at the tire display and read computer cartons that offer enough information and promise to tempt shoppers to place a thousand-dollar item into the shopping cart. I watched record store clerks rip compact discs out of their long plastic and cardboard boxes, so that music buyers wouldn't feel guilty about overpackaging.

But these research expeditions had a way of degenerating into acquisition. It would start with staple items. "I need that," I would think, grabbing a box of crackers or a few bottles of seltzer. Then after pondering the label and jar design of a new brand of spaghetti sauce, I'd find myself thinking it looked pretty good and putting a jar or two into the cart. I knew that what I was buying was mostly graphic design. Nevertheless, I granted it the ultimate critical accolade: I bought it. The same thing happened with cleansers and CDs. Now the book is done, and my house is filled with home-style pasta sauce, caustic chemicals, and the sound of music.

I hasten to add that in only a few cases did I purchase the object because I wanted to have the package. In nearly every case, my buying came the normal way — I desired the product. I know that in nearly every case, the packaging is what convinced me to buy. I even know how it did so. Despite my awareness, I was no less susceptible.

This ability of packaging to bypass the intellect and induce a consuming forgetfulness is what makes it so effective. Although packaging pervades daily life and is found in every nook and cranny of the home and workplace, it flies beneath nearly everyone's analytic radar. It only comes to the fore when there's a problem. People think about packaging when they have trouble getting it open, or when it's empty and it contributes to litter or overflowing landfills. But when packaging is working well, people rarely think about it apart from the product it contains.

Packages understand people much better than people understand packages. *The Total Package* is about how modern packaging developed and how it works now. It considers packaging not primarily as an industry or a cluster of technologies but as an important and often unacknowledged part of modern life. Its premise is that people's relationships with packages are a good deal more complex than they realize.

It is not an apology for or a condemnation of packaging, both of which would be irrelevant. Contemporary civilization could not get rid of packaging any more than the chicken could get rid of the egg.

Packaging is part of who we are. It is sometimes wasteful, sometimes misleading. But it is also clever, inventive, colorful, and full of life. And packaging is indispensable. It allows food to be distributed more widely and more efficiently. It helps make consumption predictable and has thus been essential to the development of many industries. Millions of people work in jobs that packaging helps make possible (and many of them probably think that packaging is useless).

People's relationships with packages are intimate and emotional. Packaging mirrors its expected customers, and thus it provides an unfamiliar and provocative perspective about who we are and what we want. What follows should make you more aware of the pervasiveness and power of packages; though if my experience is any guide, it won't make you resistant to their allure.

Ready?

Let's go shopping.

THE TOTAL PACKAGE

1
What's in a package

When you put yourself behind a shopping cart, the world changes. You become an active consumer, and you are moving through environments — the supermarket, the discount store, the warehouse club, the home center — that have been made for you.

During the thirty minutes you spend on an average trip to the supermarket, about thirty thousand different products vie to win your attention and ultimately to make you believe in their promise. When the door opens, automatically, before you, you enter an arena where your emotions and your appetites are in play, and a walk down the aisle is an exercise in self-definition. Are you a good parent, a good provider? Do you have time to do all you think you should, and would you be interested in a shortcut? Are you worried about your health and that of those you love? Do you care about the environment? Do you appreciate the finer things in life? Is your life what you would like it to be? Are you enjoying what you've accomplished? Wouldn't you really like something chocolate?

Few experiences in contemporary life offer the visual intensity of a Safeway, a Krogers, a Pathmark, or a Piggly Wiggly. No marketplace in the world — not Marrakesh or Calcutta or Hong Kong — offers so many different goods with such focused salesman-

1

ship as your neighborhood supermarket, where you're exposed to a thousand different products a minute. No wonder it's tiring to shop.

There are, however, some major differences between the supermarket and a traditional marketplace. The cacophony of a traditional market has given way to programmed, innocuous music, punctuated by enthusiastically intoned commercials. A stroll through a traditional market offers an array of sensuous aromas; if you are conscious of smelling something in a supermarket, there is a problem. The life and death matter of eating, expressed in traditional markets by the sale of vegetables with stems and roots and by hanging animal carcasses, is purged from the supermarket, where food is processed somewhere else, or at least trimmed out of sight.

But the most fundamental difference between a traditional market and the places through which you push your cart is that in a modern retail setting nearly all the selling is done without people. The product is totally dissociated from the personality of any particular person selling it — with the possible exception of those who appear in its advertising. The supermarket purges sociability, which slows down sales. It allows manufacturers to control the way they present their products to the world. It replaces people with packages.

Packages are an inescapable part of modern life. They are omnipresent and invisible, deplored and ignored. During most of your waking moments, there are one or more packages within your field of vision. Packages are so ubiquitous that they slip beneath conscious notice, though many packages are designed so that people will respond to them even if they're not paying attention.

Once you begin pushing the shopping cart, it matters little whether you are in a supermarket, a discount store, or a warehouse club. The important thing is that you are among packages: expressive packages intended to engage your emotions, ingenious packages that make a product useful, informative packages that help you understand what you want and what you're getting. Historically, packages are what made self-service retailing possible, and in turn such stores increased the number and variety of items people buy. Now a world without packages is unimaginable.

Packages lead multiple lives. They preserve and protect, allowing people to make use of things that were produced far away, or a while ago. And they are potently expressive. They assure that an item arrives unspoiled, and they help those who use the item feel good about it.

We share our homes with hundreds of packages, mostly in the bathroom and kitchen, the most intimate, body-centered rooms of the house. Some packages — a perfume flacon, a ketchup bottle, a candy wrapper, a beer can — serve as permanent landmarks in people's lives that outlast homes, careers, or spouses. But packages embody change, not just in their age-old promise that their contents are new and improved, but in their attempt to respond to changing tastes and achieve new standards of convenience. Packages record changing hairstyles and changing lifestyles. Even social policy issues are reflected. Nearly unopenable tamperproof seals and other forms of closures testify to the fragility of the social contract, and the susceptibility of the great mass of people to the destructive acts of a very few. It was a mark of rising environmental consciousness when containers recently began to make a novel promise: "less packaging."

For manufacturers, packaging is the crucial final payoff to a marketing campaign. Sophisticated packaging is one of the chief ways people find the confidence to buy. It can also give a powerful image to products and commodities that are in themselves characterless. In many cases, the shopper has been prepared for the shopping experience by lush, colorful print advertisements, thirty-second television minidramas, radio jingles, and coupon promotions. But the package makes the final sales pitch, seals the commitment, and gets itself placed in the shopping cart. Advertising leads consumers into temptation. Packaging *is* the temptation. In many cases it is what makes the product possible.

But the package is also useful to the shopper. It is a tool for simplifying and speeding decisions. Packages promise, and usually deliver, predictability. One reason you don't think about packages is

3

that you don't need to. The candy bar, the aspirin, the baking powder, or the beer in the old familiar package may, at times, be touted as new and improved, but it will rarely be very different.

You put the package into your cart, or not, usually without really having focused on the particular product or its many alternatives. But sometimes you do examine the package. You read the label carefully, looking at what the product promises, what it contains, what it warns. You might even look at the package itself and judge whether it will, for example, reseal to keep a product fresh. You might consider how a cosmetic container will look on your dressing table, or you might think about whether someone might have tampered with it or whether it can be easily recycled. The possibility of such scrutiny is one of the things that make each detail of the package so important.

The environment through which you push your shopping cart is extraordinary because of the amount of attention that has been paid to the packages that line the shelves. Most contemporary environments are landscapes of inattention. In housing developments, malls, highways, office buildings, even furniture, design ideas are few and spread very thin. At the supermarket, each box and jar, stand-up pouch and squeeze bottle, each can and bag and tube and spray has been very carefully considered. Designers have worked and reworked the design on their computers and tested mock-ups on the store shelves. Refinements are measured in millimeters.

All sorts of retail establishments have been redefined by packaging. Drugs and cosmetics were among the earliest packaged products, and most drugstores now resemble small supermarkets. Liquor makers use packaging to add a veneer of style to the intrinsic allure of intoxication, and some sell their bottle rather than the drink. It is no accident that vodka, the most characterless of spirits, has the highest-profile packages. The local gas station sells sandwiches and soft drinks rather than tires and motor oil, and in turn, automotive products have been attractively repackaged for sales at supermarkets, warehouse clubs, and home centers.

With its thousands of images and messages, the supermarket is as visually dense, if not as beautiful, as a Gothic cathedral. It is as complex and as predatory as a tropical rain forest. It is more than

4

a person can possibly take in during an ordinary half-hour shopping trip. No wonder a significant percentage of people who need to wear eyeglasses don't wear them when they're shopping, and some researchers have spoken of the trancelike state that pushing a cart through this environment induces. The paradox here is that the visual intensity that overwhelms shoppers is precisely the thing that makes the design of packages so crucial. Just because you're not looking at a package doesn't mean you don't see it. Most of the time, you see far more than a container and a label. You see a personality, an attitude toward life, perhaps even a set of beliefs.

The shopper's encounter with the product on the shelf is, however, only the beginning of the emotional life cycle of the package. The package is very important in the moment when the shopper recognizes it either as an old friend or a new temptation. Once the product is brought home, the package seems to disappear, as the quality or usefulness of the product it contains becomes paramount. But in fact, many packages are still selling even at home, enticing those who have bought them to take them out of the cupboard, the closet, or the refrigerator and consume their contents. Then once the product has been used up, and the package is empty, it becomes suddenly visible once more. This time, though, it is trash that must be discarded or recycled. This instant of disposal is the time when people are most aware of packages. It is a negative moment, like the end of a love affair, and what's left seems to be a horrid waste.

The forces driving package design are not primarily aesthetic. Market researchers have conducted surveys of consumer wants and needs, and consultants have studied photographs of families' kitchen cupboards and medicine chests to get a sense of how products are used. Test subjects have been tied into pieces of heavy apparatus that measure their eye movement, their blood pressure or body temperature, when subjected to different packages. Psychologists get people to talk about the packages in order to get a sense of their innermost feelings about what they want. Government regulators and private health and safety advocates worry over package design and try to make it truthful. Stock-market analysts worry about how companies are managing their "brand equity," that combination of perceived value and consumer loyalty that is expressed in advertising but embodied in packaging. The retailer is paying attention to the packages in order to weed out the ones that don't sell or aren't sufficiently profitable. The use of supermarket scanners generates information on the profitability of every cubic inch of the store. Space on the supermarket shelf is some of the most valuable real estate in the world, and there are always plenty of new packaged products vying for display.

Packaging performs a series of disparate tasks. It protects its contents from contamination and spoilage. It makes it easier to transport and store goods. It provides uniform measuring of contents. By allowing brands to be created and standardized, it makes advertising meaningful and large-scale distribution possible. Special kinds of packages, with dispensing caps, sprays, and other convenience features, make products more usable. Packages serve as symbols both of their contents and of a way of life. And just as they can very powerfully communicate the satisfaction a product offers, they are equally potent symbols of wastefulness once the product is gone.

Most people use dozens of packages each day and discard hundreds of them each year. The growth of mandatory recycling programs has made people increasingly aware of packages, which account in the United States for about forty-three million tons, or just under 30 percent of all refuse discarded. While forty-three million tons of stuff is hardly insignificant, repeated surveys have shown that the public perceives that far more than 30 percent — indeed, nearly all — their garbage consists of packaging. This perception creates a political problem for the packaging industry, but it also demonstrates the power of packaging. It is symbolic. It creates an emotional relationship. Bones and wasted food (13 million tons), grass clippings and yard waste (thirty-one million tons), or even magazines and newspapers (fourteen million tons) do not feel as wasteful as empty vessels that once contained so much promise.

Packaging is a cultural phenomenon, which means that it works differently in different cultures. The United States has been a good market for packages since it was first settled and has been an important innovator of packaging technology and culture. Moreover, American packaging is part of an international culture of modernity and consumption. At its deepest level, the culture of American packaging deals with the issue of surviving among strangers in a new world. This is an emotion with which anyone who has been touched by modernity can identify. In lives buffeted by change, people seek the safety and reassurance that packaged products offer. American packaging, which has always sought to appeal to large numbers of diverse people, travels better than that of most other cultures.

But the similar appearance of supermarkets throughout the world should not be interpreted as the evidence of a single, global consumer culture. In fact, most companies that do business internationally redesign their packages for each market. This is done partly to satisfy local regulations and adapt to available products and technologies. But the principal reason is that people in different places have different expectations and make different uses of packaging.

The United States and Japan, the world's two leading industrial powers, have almost opposite approaches to packaging. Japan's is far more elaborate than America's, and it is shaped by rituals of respect and centuries-old traditions of wrapping and presentation. Packaging is explicitly recognized as an expression of culture in Japan and largely ignored in America. Japanese packaging is designed to be appreciated; American packaging is calculated to be unthinkingly accepted.

Foods that only Japanese eat — even relatively humble ones like refrigerated prepared fish cakes — have wrappings that resemble handmade paper or leaves. Even modestly priced refrigerated fish cakes have beautiful wrappings in which traditional design accommodates a scannable bar code. Such products look Japanese and are unambiguously intended to do so. Products that are foreign, such as coffee, look foreign, even to the point of having only Roman lettering and no Japanese lettering on the can. American and European companies are sometimes able to sell their packages in Japan virtually unchanged, because their foreignness is part of their selling power. But Japanese exporters hire designers in each country to repackage their products. Americans — whose culture is defined

7

not by refinements and distinctions but by inclusiveness — want to think about the product itself, not its cultural origins.

We speak glibly about global villages and international markets, but problems with packages reveal some unexpected cultural boundaries. Why are Canadians willing to drink milk out of flexible plastic pouches that fit into reusable plastic holders, while residents of the United States are believed to be so resistant to the idea that they have not even been given the opportunity to do so? Why do Japanese consumers prefer packages that contain two tennis balls and view the standard U.S. pack of three to be cheap and undesirable? Why do Germans insist on highly detailed technical specifications on packages of videotape, while Americans don't? Why do Swedes think that blue is masculine, while the Dutch see the color as feminine? The answers lie in unquestioned habits and deep-seated imagery, a culture of containing, adorning, and understanding that no sharp marketer can change overnight.

There is probably no other field in which designs that are almost a century old — Wrigley's gum, Campbell's soup, Hershey's chocolate bar — remain in production only subtly changed and are understood to be extremely valuable corporate assets. Yet the culture of packaging, defined by what people are buying and selling every day, keeps evolving, and the role nostalgia plays is very small.

For example, the tall, glass Heinz ketchup bottle has helped define the American refrigerator skyline for most of the twentieth century (even though it is generally unnecessary to refrigerate ketchup). Moreover, it provides the tables of diners and coffee shops with a vertical accent and a token of hospitality, the same qualities projected by candles and vases of flowers in more upscale eateries. The bottle has remained a fixture of American life, even though it has always been a nuisance to pour the thick ketchup through the little hole. It seemed not to matter that you have to shake and shake the bottle, impotently, until far too much ketchup comes out in one great scarlet plop. Heinz experimented for years with wide-necked jars and other sorts of bottles, but they never caught on.

Then in 1992 a survey of consumers indicated that more

Americans believed that the plastic squeeze bottle is a better package for ketchup than the glass bottle. The survey did not offer any explanations for this change of preference, which has been evolving for many years as older people for whom the tall bottle is an icon became a less important part of the sample. Could it be that the difficulty of using the tall bottle suddenly became evident to those born after 1960? Perhaps the tall bottle holds too little ketchup. There is a clear trend toward buying things in larger containers, in part because lightweight plastics have made them less costly for manufacturers to ship and easier for consumers to use. This has happened even as the number of people in an average American household has been getting smaller. But houses, like packages, have been getting larger. Culture moves in mysterious ways.

The tall ketchup bottle is still preferred by almost half of consumers, so it is not going to disappear anytime soon. And the squeeze bottle does contain visual echoes of the old bottle. It is certainly not a radical departure. In Japan, ketchup and mayonnaise are sold in cellophane-wrapped plastic bladders that would certainly send Americans into severe culture shock. Still, the tall bottle's loss of absolute authority is a significant change. And its ultimate disappearance would represent a larger change in most people's visual environment than would the razing of nearly any landmark building.

But although some package designs are pleasantly evocative of another time, and a few appear to be unchanging icons in a turbulent world, the reason they still exist is because they still work. Inertia has historically played a role in creating commercial icons. Until quite recently, it was time-consuming and expensive to make new printing plates or to vary the shape or material of a container. Now computerized graphics and rapidly developing technology in the package-manufacturing industries make a packaging change easier than in the past, and a lot cheaper to change than advertising, which seems a far more evanescent medium. There is no constituency of curators or preservationists to protect the endangered package. If a gum wrapper manages to survive nearly unchanged for ninety years, it's not because any expert has determined that it is an important cultural expression. Rather, it's because it still helps sell a lot of gum.

9

*   *   *

So far, we've been discussing packaging in its most literal sense: designed containers that protect and promote products. Such containers have served as the models for larger types of packaging, such as chain restaurants, supermarkets, theme parks, and festival marketplaces.

This limited, but still vast, domain of packaging is what this book is about. Still, it is impossible to ignore a broader conception of packaging that is one of the preoccupations of our time. This concerns the ways in which people construct and present their personalities, the ways in which ideas are presented and diffused, the ways in which political candidates are selected and public policies formulated. We must all worry about packaging ourselves and everything we do, because we believe that nobody has time to really pay attention.

Packaging strives at once to offer excitement and reassurance. It promises something newer and better, but not necessarily different. When we talk about a tourist destination, or even a presidential contender, being packaged, that's not really a metaphor. The same projection of intensified ordinariness, the same combination of titillation and reassurance, are used for laundry detergents, theme parks, and candidates alike.

The imperative to package is unavoidable in a society in which people have been encouraged to see themselves as consumers not merely of toothpaste and automobiles, but of such imponderables as lifestyle, government, and health. The marketplace of ideas is not an agora, where people haggle, posture, clash, and come to terms with one another. Rather, it has become a supermarket, where values, aspirations, dreams, and predictions are presented with great sophistication. The individual can choose to buy them, or leave them on the shelf.

In such a packaged culture, the consumer seems to be king. But people cannot be consumers all the time. If nothing else, they must do something to earn the money that allows them to consume. This, in turn, pressures people to package themselves in order to survive. The early 1990s brought economic recession and shrinking

10

opportunities to all the countries of the developed world. Like products fighting for their space on the shelf, individuals have had to re-create, or at least re-present, themselves in order to seem both desirable and safe. Moreover, many jobs have been reconceived to depersonalize individuals and to make them part of a packaged service experience.

These phenomena have their own history. For decades, people have spoken of writing resumes in order to package themselves for a specific opportunity. Thomas J. Watson Jr., longtime chairman of IBM, justified his company's famously conservative and inflexible dress code — dark suits, white shirts, and rep ties for all male employees — as "self-packaging," analogous to the celebrated product design, corporate imagery, and packaging done for the company by Elliot Noyes and Paul Rand. You can question whether IBM's employees were packaging themselves or forced into a box by their employer. Still, anyone who has ever dressed for success was doing a packaging job.

Since the 1950s, there have been discussions of packaging a candidate to respond to what voters are telling the pollsters who perform the same tasks as market researchers do for soap or shampoo. More recently, such discussions have dominated American political journalism. The packaged candidate, so he and his handlers hope, projects a message that, like a Diet Pepsi, is stimulating without being threatening. Like a Weight Watchers frozen dessert bar, the candidate's contradictions must be glazed over and, ultimately, comforting. Aspects of the candidate that are confusing or viewed as extraneous are removed, just as stems and sinew are removed from packaged foods. The package is intended to protect the candidate; dirt won't stick. The candidate is uncontaminated, though at a slight remove from the consumer-voter.

People profess to be troubled by this sort of packaging. When we say a person or an experience is "packaged," we are complaining of a sense of excessive calculation and a lack of authenticity. Such a fear of unreality is at least a century old; it arose along with industrialization and rapid communication. Now that the world is more competitive, and we all believe we have less time to consider things, the craft of being instantaneously appealing has

11

taken on more and more importance. We might say, cynically, that the person who appears "packaged" simply doesn't have good packaging.

Still, the sense of uneasiness about encountering packaged people in a packaged world is real, and it shouldn't be dismissed. Indeed, it is a theme of contemporary life, equally evident in politics, entertainment, and the supermarket. Moreover, public uneasiness about the phenomenon of packaging is compounded by confusion over a loss of iconic packages and personalities.

Producers of packaged products have probably never been as nervous as they became during the first half of the 1990s. Many of the world's most famous brands were involved in the merger mania of the 1980s, which produced debt-ridden companies that couldn't afford to wait for results either from their managers or their marketing strategies. At the same time, the feeling was that it was far too risky to produce something really new. The characteristic response was the line extension — "dry" beer, "lite" mayonnaise, "ultra" detergent. New packages have been appearing at a rapid pace, only to be changed whenever a manager gets nervous or a retailer loses patience.

The same skittishness is evident in the projection of public personalities as the clear, if synthetic, images of a few decades ago have lost their sharpness and broken into a spectrum of weaker, reflected apparitions. Marilyn Monroe, for example, had an image that was, Jayne Mansfield notwithstanding, unique and well defined. She was luscious as a Hershey's bar, shapely as a Coke bottle. But in a world where Coke can be sugar free, caffeine free, and cherry flavored (and Pepsi can be clear!), just one image isn't enough for a superstar. Madonna is available as Marilyn or as a brunette, a Catholic schoolgirl, or a bondage devotee. Who knows what brand extension will come next? Likewise, John F. Kennedy and Elvis Presley had clear, carefully projected images. But Bill Clinton is defined largely by evoking memories of both. As our commercial civilization seems to have lost the power to amuse or convince us in new and exciting ways, formerly potent packages are recycled and devalued. That has left the door open for such phenomena as generic cigarettes, President's Choice cola, and H. Ross Perot.

12

This cultural and personal packaging both fascinates and infuriates. There is something liberating in its promise of aggressive self-creation, and something terrifying in its implication that everything must be subject to the ruthless discipline of the marketplace. People are at once passive consumers of their culture and aggressive packagers of themselves, which can be a stressful and lonely combination.

The Total Package is obviously a response to this time and this set of feelings. For the most part, though, these existential questions of packaging will lurk somewhere in the background, while the text explores the history of protective, standardized, informative, and emotion-laden containers and the forces that shape them today.

Even once you decide to view packaging in the relatively narrow sense of communicative containers, defining what you mean is still difficult. Most industry sources begin with the notion that a package is a container and then add such attributes as suitability for the manufacturing process, ease of shipping, attractiveness of display, and convenience for the ultimate user. Some experts list as many as fifty key characteristics of packaging, a seeming precision that results in curious ambiguities. Most statistics on packaging exclude such elements as shopping bags and gift wrappings because they are unrelated to a specific product, although they do have emotional and promotional characteristics and, like a store's interior design, serve as part of an overall retailing package.

The statistics usually include corrugated paperboard shipping cartons, which add a lot of weight to the subject but, until recently, not much spirit. Environmentalists like to include the tonnage of shipping containers in the statistics of garbage generated, because it helps to suggest that each man, woman, and child is personally discarding half a ton a year in packages. (The figure is closer to 370 pounds.) The packaging industry likes to point to shipping cartons because more than half of them are recycled, greatly enhancing the industry's environmental record.

Such cartons are an important component of packaging because they protect products during their periods of roughest handling as they are transported and distributed. Strong packaging at this stage of the product's life can allow point-of-sale packages to be less sturdy. And because most shipping cartons are recycled by industry, while individual packages go in the household trash, it's simply more efficient to have strong cartons and weak consumer packages.

If you view packaging as a system that links production and consumption, shipping cartons are an extremely important part of it. Traditionally, consumers have had little awareness of such containers because they were kept out of the shopper's way. Recently, however, largely because of the influence of mass, self-service retailers, shipping cartons for many grocery products have been designed to serve as lively, colorful point-of-sale displays, while others containing appliances and machinery have become attractive and often contain detailed information about the use of the product, its specifications, energy use, warranty, and other items of great interest to the buyer. The first kind of carton serves as a substitute for in-store promotion and does not leave the store with the buyer. The second kind of carton replaces the expertise that was once expected of a salesman, as more and more products are sold in self-service environments. It is also likely to become part of the buyer's life, filling up home storage space until the warranty has expired. Belatedly, shipping containers are coming out of the back room to take on the role of selling and informing that we tend to expect of a full-fledged package.

One way to view the carton is as one layer of a many-layered system of packaging. The extreme outer layer might be a wooden pallet on which cartons have been secured with strong plastic film, so that they can easily be moved with a forklift or perhaps by a robot. Inside the shipping carton are typically several more layers of packaging. The package the consumer purchases might itself be wrapped in clear film, either to keep moisture out or, in the case of food, to keep it in. The box itself is likely to have many layers, each of which performs a specialized task of protection or graphic expression. The image might consist of multiple layers with foil, plastic, paper, color, and even a layer to protect the graphic from being harmed. It may sound odd to consider ink as a layer, but it can actually be just about as thick as the other "solid" layers that make up a package. In

14

contemporary packages, complex multilayer components can be extremely thin. Some potato chip bags fit as many as nineteen layers, each of which has a function, in much less than a thousandth of an inch.

The layering that's typical of packaging is analogous to the layering of clothing. Some items of clothing are ornamental, others provide warmth, and still others protect the body from being irritated by other layers of clothing and protect the outer layers from perspiration or other bodily phenomena that shouldn't be visible. The layering of clothing is complex to explain, but intuitive to do. The layering of packaging is equally sensible, but sometimes controversial because it can make recycling more difficult.

In fact, the layering of packaging is not particularly new. For most of two centuries, canning has depended on tinplate — iron, then steel coated with a thin layer of tin — to prevent the metal from rusting. From the early nineteenth century, chocolate and tea were sold in multilayered packages, and cigarette packs have for a century had a foil inner wrap and a paper outer wrap and since the 1940s a transparent outer wrap.

The layers of a package can be a lure, a reassurance, or an irritation. In the late 1980s, when Haagen-Dazs, a luxury ice-cream brand, introduced ice-cream bars, the package consisted of wrapped bars set in a plastic tray within a box that was displayed in the supermarket freezer. The tray was intended not only to present an opulent appearance for the expensive dessert, but also to protect the bars from damage in shipping. But market surveys revealed that the tray was viewed very negatively by those who had bought the product. They didn't mind paying extra for a better ice cream, they said, but they weren't going to pay extra for a big, useless tray they had to throw away. People will often tell market researchers that they'll pay more for products that are in packages they believe express high quality. But the minute they think they are paying extra for the package, they rebel. Haagen-Dazs designed a stronger outer box to compensate for the stiffness that had been provided by the tray and got rid of what consumers viewed as extraneous.

Recently, the editor in chief of the trade magazine *Packaging* provoked a rare philosophical debate among his readers when he suggested that the plastic "jewel box" in which compact discs are sold is not a package. His argument was that because the holder remains with the disc throughout its lifetime, it is an inseparable part of the product. His readers replied that the jewel box is like countless other packages that contain the product throughout its period of usefulness. The only difference is that the CD, in theory at least, doesn't wear out or get used up. Nobody ever debated whether a long-playing album's cover was a package because it provided such a large area for the graphic salesmanship that is so important a part of packaging.

This minor controversy suggests a possible distinction: a package consists of the content of the product, plus all of the other elements that are needed to protect the contents when it is being shipped, keep track of inventory, and entice the customer at the point of purchase. Everything that's not the product can then be termed the packaging.

There's only one problem: as a practical matter it's almost impossible to define what's not the product. The great majority of products are not simply contained commodities. They are, rather, combinations of materials, information, and containers, each of which is essential to the usefulness of the total package. You literally can't separate the product from the packaging of lipstick, a book of matches, or a can of shaving cream. Packages create new necessities: such goods as shampoos, breakfast cereals, condensed soups, multipurpose cleaners, pancake mixes, frozen dinners, and countless other products literally did not exist until they were packaged. Even the shipping container, which you usually don't see on the shelf, is part of the product: if what you are selling is a beer from the Rockies, it doesn't exist unless you can get it safely from Colorado to all the places where people buy it.

And even those things that seem most extraneous to the physical package — the sparkling pictures and other marketing lures — are also part of the product. In order to be able to produce something, you have to have a mechanism to produce and control its consumption. Packaging links industry to dreams and helps keep them in sync. When production doesn't match promises, and aspirations don't stay in tune with what can be delivered, modernity grinds to a halt. Communism collapses, and affluent societies look nervously toward the millennium. Suffice it to say that the distinction between the product and its packaging, while intuitively obvious — and probably valid less than a century ago — is no longer sustainable. We've come too far for that.

If layering is an essential characteristic of packaging, so is bundling. The earliest dictionary definition of "package" described it as something that has been packed, typically to be shipped, as with cloth, or to be sold by an itinerant peddler. Such a package could consist of disparate things, and indeed, some early citations of the word speak of smugglers' packages that contained things other than what was claimed. A package can contain many different things, but they are concealed behind a promise.

What's important about the package is that somebody put it together for the ultimate user. If you go to a grocer and ask for a pound of flour, which he puts in a paper bag and hands to you, that's not really a package. But if he had filled pound bags in advance and labeled them, the combination of containers with preparation and information makes them packages.

You can also come at a definition of "package" from the point of view of the buyer rather than the seller and argue that a package is an indivisible unit of consumption. The grocer with a flour barrel could sell you half a pound of flour if you asked, but at the supermarket, you have to take the size package that's available. Alternatively, you might buy the flour along with a number of other measured ingredients in the form of a box of cake mix, complete with a color picture of the ultimate cake on the front of the box. And though it is unlikely that a baker will sell half a loaf of bread, he might as a favor for a loyal or convincing buyer. But once the bread is put into a bag and sent to the supermarket, that possibility no longer exists. Packages replace the indeterminacy of human relationships with quantities and emotional messages, both of which have been carefully measured.

This suggests a more philosophical definition of the package as a tool of expression and knowledge. Packaging helps people

17

know and decide things quickly, a task it accomplishes by a combination of display and concealment. Display is the most obvious element of packaging. But what the package shows is tightly edited to focus on the satisfaction that the product will provide. Consumers are interested in whether the cleaner will remove mildew from the shower stall or whether the frozen dinner promises to be tasty. A few may look carefully at ingredients and nutritional information, which are required by law. But the promise of packaging is that you don't have to worry about the process that brings a product into being. You can make a good decision without even having to think about it.

Expressed in this way, packaging has something menacing about it. It implies a denigration of intellect, from which follows a loss of control and, hence, a loss of human dignity. These are serious issues. But how much do you really want to control? Do you really want to worry about the chemistry of killing mildew, one of the earth's more enigmatic life-forms? Do you really have the energy to make lasagna — now?

There is, however, a better argument for the tightly edited communication of packaging than simple laziness. Just because much of people's response to packages is emotional and subconscious, that doesn't mean that people are gullible or powerless. Rather, it is the kind of rapid, unreflective communication of information that allows, for example, the pilot of a jet fighter plane to guide an object moving at supersonic speeds as if it were his own body. The point of being precise and evocative in your packaging design, then, is not to manipulate but to communicate so effectively that the shopper will understand immediately. From this point of view, such instant recognition and understanding reveals the effective package to be an advanced technology and a model for the many complex information interfaces — from industrial control systems to five-hundred-channel cable menus — that are part of contemporary life.

It's not necessary to be a high-technology jockey to find a way of understanding packages as useful tools that actually enhance human power.

The protective function of packaging is often taken for granted, but it is in fact a very powerful set of technologies with life and death consequences. Much of the hunger in the world is caused not by the inability to produce food, but by failures in getting food to people before it is spoiled. The U.S. Department of Agriculture estimates that half of the food produced for sale in the world (excluding what is grown by subsistence farmers for their own use) rots before it can be eaten. United Nations estimates in the late 1980s stated that 70 percent of the food grown in India is wasted because of spoilage, while the comparable figure for the United States was 17 percent. (Other estimates, using different methods, place the U.S. figure as low as 4 percent.) Thus, while American agriculture uses methods that make extremely intensive use of energy and capital investment, processing and packaging contribute substantially to the efficiency of the entire food production system and are two reasons American food costs are low compared with most of the rest of the world. Canned and frozen vegetables lack the sensual quality of fresh produce in a traditional marketplace, but they do contribute greatly to the ability of developed, industrialized nations to nourish, indeed overnourish, their populations.

Moreover, packaging technologies also enable many more people to have a varied diet throughout the year and provide sources of important vitamins and minerals. There's no doubt that fruits and vegetables in season taste best, but most people are used to the idea that there will be tomato sauce even when there are no good tomatoes in the market, and that frozen orange juice is available no matter what's happening in the citrus groves. Improved transportation and storage methods have, of course, made it possible, for example, to sell Chilean peaches in New York in winter, thus freeing people from having to use the canned variety. But in most agricultural industries, the fresh market is only a small portion of the whole, a welcome dividend of the principal processed markets.

In 1953 the food writer Poppy Cannon described the can opener as "That *open sesame* to wealth and freedom. . . . Freedom from tedium, space, work and your own inexperience." She could scarcely have imagined how large a role packaged and processed foods would play in the emancipation of the woman from the

ALBUQUERQUE ACADEMY
LIBRARY

kitchen. As recently as the late 1970s, according to an estimate by Andersen Consulting, Americans — mostly wives — spent nearly three and a half hours each day preparing three meals for their households. By the early 1990s, the figure had dropped to thirty minutes. To be sure, the reduced cooking time is not entirely accounted for by packaged foods, only the largest share. Fast-food restaurants — themselves a packaged phenomenon — have replaced some of the lost cooking time, as has the sale of prepared foods by restaurants, take-out businesses, and supermarkets. Such prepared food is a fast-growing profit center for supermarkets, which are giving it a branded identity and spurring container improvements that are characteristic of a packaged product.

You can argue that such a severe reduction of food preparation time — and the concomitant devaluation of the dinner table as the place where family ties are forged — isn't progress at all. The two-income family is a liberation for some, a trap for others. Packaging has surely helped shape society's expectations of ever-increasing consumption that requires households to have two paychecks to keep up. But if packaging has played a role in creating this problem, it has also helped people cope with it.

All packages contain some measure of information. Even if all the package tells you is what it contains and how much of it, that's useful. Moreover, the information can generally be trusted. Adulteration of packaged products is rare. Short-weighting of contents is far more frequent, but it is seldom dramatic. And while neighborhood grocers were once famous for their heavy fingers on the scale, shoppers generally believe that brand-name packaged products are precisely what they say they are.

Moreover, shopping by the list is the exception rather than the rule. For example, grocery shoppers may have a general idea of what they are looking for, but far more often than not, they go to the supermarket, walk down the aisle, and let the packages speak to them. By surrendering in this way to the irrational, they play into the hands of food manufacturers and retailers. Such unplanned shopping leads to far higher grocery purchases than does a list based on rigorous analysis of household needs. But such a focused, rational approach entails a level of effort that many prefer not to make. On the store's shelves is a vast array of appetizing suggestions, and on the packages' backs, there is a library's worth of recipes.

The communicative power of packages is so strong that consumer advocates have fought, with some success, to reshape them as educational tools. Now it's possible to look at a can of peaches, for example, and learn precisely what quantities of nutrients it contains and how much of your recommended daily intake it provides. There might even be some suggestions about what else you ought to eat to achieve a healthy diet. A fresh peach may be juicier, but it does not begin to tell you so much. Some packages, such as cigarettes and liquor bottles, are required by law to say that their contents are harmful, though they tend to do it in condensed light-face type that is difficult to read. The same is true for disclosure of contents or processes that most people would prefer not to know about. "From Concentrate" on fruit juice containers and "Contains Sulfites" on wine bottles are among the skinniest words in the language.

It's not clear whether packages will ever really teach people about nutrition. Their principal focus will probably always be to convince buyers that what's inside is satisfying.

However, the informational content of nonfood packaging has become extremely important. Many products that were once sold as part of a service, such as tires, have made it to self-service outlets. Some tires are now shrink-wrapped with data on tread life, stopping distances, and other important information visible to the buyer. Computers and other electronic equipment come in cartons that illustrate and describe the equipment inside and, in some cases, also contain instructions about how to set up the equipment and how to recycle the packing materials. Even such big-ticket items are marketed under the assumption that the package must answer potential buyers' questions and assuage their concerns. The package that offers clear, useful graphics and writing offers the best service a customer is likely to receive. Much of the time it is the only service.

This concept is hardly new. "The display container is as much of a salesman as any flesh-and-blood clerk, and often more," wrote Richard B. Franken and Carroll B. Larabee in their 1928 book *Packages That Sell*, "for it works night and day for one product and emphasizes only those sales arguments which the manufacturer knows are best."

The triumph of packaging is an aspect of the larger stories of industrialization, control technologies, and the consumer society. The substitution of packages for salesclerks began after machines started displacing craftsmen and before office clerks gave way to computers. But as a historical phenomenon, it has received a good deal less study and attention than either of those processes. Today, the salesclerk is virtually extinct, and aside from those who used to be salesclerks, few have really noticed. Indeed, many surveys indicate that in most situations, people believe in packages and their own judgment far more than they trust salespeople.

And salesclerks aren't the only people packages try to replace; they seek to overcome the advice of mothers, friends, neighbors, and other sources of authority. Packages have personality. They create confidence and trust. They spark fantasies. They move the goods quickly.

Large-scale self-service retailing is at most seventy years old, and it only became universal after World War II. But it vastly increased the velocity with which people consume things. Traditional, personal-service marketplaces are picturesque and make desirable tourist attractions, but most Americans, and rapidly increasing numbers of people elsewhere, find it far too difficult to shop in such places for life's necessities. Imagine how much more slowly you would acquire things if every purchase required that you personally ask somebody for it, hear an opinion on whether you were making the right choice, and then wait while the item was retrieved from stock. How many fewer items would be purchased on impulse if you couldn't see and then grab the package off a shelf?

Today, packaging is reflecting a broad technological and cultural trend by increasing its information content and decreasing its material content. That may not seem evident when you glance at a landfill and see all the bright alluring colors you remember from the supermarket. Nevertheless, the amount of material used in most packaging is being rapidly diminished, as cans and bottles have become thinner and in some cases given way to flexible pouches. This is an environmental approach that manufacturers have embraced with some enthusiasm, because it also helps them to save money. William Rathje, the University of Arizona archaeologist who digs up landfills, has confirmed that today's trash is more crushable, and hence more compact, than that of a decade and more ago.

At the same time, those who design and make packages try to look ahead to new retail environments. One of these might be a store full of smart packages, in which touching the product can produce an advertisement on your shopping cart's video screen, a pharmaceutical container can announce when to take the medicine and how much to take, supermarket computers can generate customized cookbooks, or products that need assembly might incorporate three-dimensional holographic instructions.

Packagers are looking at the prospect of life after shopping carts. Technology may soon make it cheaper for retailers to maintain superwarehouses and provide home delivery of everything. Although computer-based retailing has been disappointing, and television shopping is just starting to catch on, shopping for everything through computers and video will require new technologies and new ideas. The virtual supermarket will require electronic packages — or more accurately a new layer of information and enticement over the protective layers that already exist. Packaging a product electronically has to be something different than creating a television commercial or a catalog illustration. Will it mean that the looks of the container itself won't matter anymore? Will packages ordered from home have to look like people's interior decor? Or will designers have to change their emphasis from closing the sale to affirming a wise choice? Stay tuned.

If some of the wilder visions are right, *The Total Package* comes at a moment just before the package disappears from everyday life. Don't count on that prediction coming true just yet.

Packaging is part of human behavior, and it is a very powerful tool for communication and understanding. That won't change even if you could throw away the container.

2 *Holding magic*

*I*n 1992 archaeologists determined that a five-thousand-year-old vessel, excavated from a mountain settlement in the Zagros Mountains of western Iran, contains residue that is the earliest chemical evidence of beer. The site where it was found, known as Godin Tepe, is on what later became the Silk Route, and it is believed to have been a Sumerian trading outpost. Packages are often associated with outposts and places where strangers come together.

The residue was found in shards from a jar that had been deeply scored on the inside. The archaeologists believe these internal crisscross grooves were functional, intended to collect the bitter-tasting sediment — calcium oxalate, or beer stone — that is an unavoidable product of the brewing process. The beer was apparently both stored and served in these vessels. Reliefs on ancient seals show drinkers sitting on chairs next to the large urns, sipping the beer through straws.

Obviously, such urns were not packages, but they were forerunners, protopackages. The package as we know it today is unquestionably a modern phenomenon that couldn't have happened without printing presses, urbanization, industrialization, railroads, and the rest of the tools of organization and communication that converged more than a century ago to change everything.

But just as there was writing before there were presses, and cities before there were industrial metropolises, there were many things that preceded the modern package and fulfilled many of the same practical and symbolic functions. It seems likely that the emotional force that packaging can possess springs from its deep cultural roots. Packages are about containing and labeling and informing and celebrating. They are about power and flattery and trying to win people's trust. They are about beauty and craftsmanship and comfort. They are about color, protection, and survival. By considering the nature of the precursors to packaging — both man-made and natural — we can possibly understand some of the hidden content of the packages that fill our stores, refrigerators, dressing tables, and medicine cabinets.

The shards found at Godin Tepe suggest that the vessel from which they came was used for storing and serving, two important roles for packaging. The urn was apparently not used for transporting the beer; barley, the chief ingredient of beer, was found on the floor of the room where the jar was found.

As it happens, the earliest chemical evidence of wine was found in a jar from the same room at Godin Tepe. It too may have been made nearby. The jar that contained the wine was different from the one that contained the beer. Unlike the beer jar, which stood upright, the wine jar was made to be stored on its side and be capped with an unfired clay stopper that, like a cork, would absorb some wine, swell, and keep out air to retard spoilage. The room was apparently some sort of distribution center, where two different kinds of containers contained two different commodities that had different functional requirements. They also had different "markets": beer was the drink of the lowlands to the west, while wine came from highlands to the north.

The most interesting connection with packaging, however, concerns language. The beer vessel was a product of the same civilization that invented cuneiform writing. The earliest character for beer, *kas*, seems to depict an urn similar to the one in which the beer was found, with lines or crosshatches inside. The archaeologists who made the discovery hypothesize that the lines represent the internal scoring of the vessel that was used to collect the residues, something that was done uniquely for beer. Another interpretation is that the lines simply indicated that the vessel was full of liquid, though nobody disagrees that this symbol does mean beer and nothing else. Thus, in writing, the product was identified with a particular container, and, symbolically, the container stood for the thing itself. This strong identification of container with contents, and the container with symbolism, suggests that the communicative dimension of packaging — which appears to be its most modern aspect — actually has very ancient roots. Indeed, you could go so far as to say that it is part of the way people think.

When you start to look at the archaeological record for the precursors to packaging, it's hard to know where to begin or end. Whether you are excavating the tomb of a king or nobleman or making your way through an ancient garbage pile, most of what you find can be considered a forerunner to packaging. Moreover, it seems clear that people were making baskets and other kinds of fiber containers thousands of years before they were able to make the hard, long-lasting objects that constitute most of the archaeological record.

Some ancient vessels were clearly used primarily for transport, while others were personal possessions and used in rituals of religion or hospitality. Some richly worked containers clearly had great religious and political significance. A mummy case is a package of sorts: it contains the human remains, shows a picture of the person, and usually contains written information that provides a social and cosmological context.

Nearly everywhere — from Greece to India to Mesoamerica — there are ancient vessels that take the shapes of humans and animals. Some of them are menacing, others erotic, and still others seem to embody a celebration of joy and beauty. The meanings and uses of such objects are complex, varied, and frequently enigmatic. Such figural vessels express the power of a container to transform and animate its contents. For example, the traditional *minkisi* of Zaire are figures of human shape that contain what are believed to be powerful substances, such as crushed bone and special herbs. The power of the *minkisi*, like the power of a cleaning agent or a perfume, is activated when an opening is made in the container. It costs something to do that.

Containers from times or places that are remote are often seen as tools for what we call magic: a quest for power and understanding through the assertion of associations between seemingly unrelated things. Contemporary packaging often suggests unlikely associations, too — between superheroes and sugary cereals, for instance. And the shelves are full of polishes, detergents, and engineered foods that claim to work like magic.

It's also worth noting that the oldest-known graphic expressions, the cave paintings of southern France and Spain, might represent a kind of magic akin to what is found on packages. The paintings, many of which show large animals pierced by spears and arrows, are believed by many to have been part of a magic of anticipation. The pictures on the wall were used to make a good hunt more likely. Consider the pictures on packaging — the "serving suggestion" picture that depicts a steaming, colorful entrée light-years removed from the frozen lump in the plastic tray you are actually buying. Recall the shampoo container on which the hair is so radiant that it creates an irresistible sexual aura, or the numerous products that have, for close to a century, been sold with the image of a healthy, contented smiling baby. We understand these images to be at odds with experience. But there is always hope; there is always magic.

Correspondences between a vessel and its contents are rarely obvious, and sometimes they might not even exist. If a vessel has clusters of grapes painted on the outside, there is good reason to guess that it might have been made for wine. A widemouthed jar with fish on the outside might have been used to transport preserved fish or the popular condiments made from fermented salted fish. But we can't often be sure that it did. Even today, pictures of

27

animals on packages are more likely to be associated with the attributes of the beast than the contents of a package. We look at a bull or an elephant on a bottle of malt liquor and know that this product is powerful and will trample the drinker into a state of quick oblivion. A whale used to be a marker of a really big box of detergent.

The archaeological record of vessels, boxes, bottles, and jars from different places and times is vast. But archaeologists, like shoppers, tend to look past the container to find something else — usually evidence about how people understood the world, how they worshiped, how they traded or ate. There is no archaeology without old containers, but very little archaeology that focuses on containers.

The best-known container of the ancient Mediterranean region is the so-called Canaanite jar, or amphora, a vessel that took on its classic form in about 1800 B.C. and was used, with relatively minor variations, for more than two thousand years. It was generally about thirty liters in capacity, with small handles at the top and a rounded bottom that, like the similar bottoms of PET (polyethylene terephthalate) plastic soft-drink bottles, helped distribute the pressures the container must withstand more evenly. These jars could be piled on their sides several layers deep in the holds of ships or attached one to each flank of a donkey. They carried wine, oil, and other liquid goods, sometimes even water. They were, in a sense, universal containers, the ancient equivalent of corrugated cartons or fifty-five-gallon oil drums. Sometimes, after their contents had been consumed, their narrow necks were cut off, and they were used for the burial of babies. Herodotus, in describing life in Egypt, said that such jars were universally used, but empty ones were rarely encountered by visitors because of a massive recycling program. He said that the leaders of cities and villages were responsible for the collection of these jars, which were then refilled with water and transferred to posts along the desert approaches to Egypt, so that long-distance travelers would not have to carry as much water as would otherwise be necessary.

In a sense, such universal containers are the opposite of packages, because their form expressed little about their contents.

Most likely, consumers would have brought their own vessels to merchants to be filled from these large jars. And in many cases, the pitchers or carafes used might, through their form or decoration, communicate something about their contents. Pieces for serving wine might have a relief of a bunch of grapes or a mythological scene associated with wine. People still use such serving vessels, probably in greater quantity than the ancients did, though we tend not to think of them as packages. And with only a few prominent exceptions, such as premium liquor bottles or perfumes, people don't display packages to impress their guests. But in daily life, packages do constitute a significant portion of the household landscape, and they replace many sorts of canisters, cruets, and other household containers that were required when products were bought in bulk rather than in packages.

The one way in which the Canaanite jar was a true precursor to the package was in its labeling. There are examples of jars in ancient Egypt stamped with the name of the reigning pharoah as early as 5,000 years ago. This could be a mark of vintage, of ownership, or even a tax stamp, but it did not reveal the contents. But by 3,300 years ago, Canaanite jars containing wine were routinely stamped with information at least as useful as that found on wine labels today. The inscriptions detailed the date of the wine, the type of grape and where it was grown, and whether the wine was sweet or dry. There is a sense in which such labeling seems rather late in coming, given that jars so resembled one another and that they were opaque and often securely stoppered. Thus, it was often impossible or at least destructive to take a peek at or a taste of what was inside.

But you can also view the markings on the Canaanite wine jar as a precocious form of branding. They imply that the product was consumed some distance from where it was produced and, more important, that those who looked at the label had a pretty good idea of what to expect even before they tasted the wine. That implies some form of standardization or quality control, and people who bought the wine probably decided what they would pay on the basis of what this label said. It was an implicit promise that, as with modern packages, sped up the buying and selling process and eliminated the waste that results from having to sample the product to make sure it's right. Information is an essential component of packaging, even though it does not always tell the whole truth.

"A package," the designer Ernest Dichter has written, "is the expression of the respect we have for the consumer."

The redefinition of people as consumers is quite a recent phenomenon. But the idea of presenting gifts in lavish or ornamental containers sometimes as a personal expression of affection, though most often as a gesture of deference to a powerful person, is both ancient and widespread. Paying tribute is often a gesture of subservience, the acceptance of another's sovereignty. Because it is done from a position of weakness, it is important that it should be done carefully, in a presentation that will not provoke any displeasure. The custom survives today in the diplomatic ritual of mutual gift giving that accompanies state visits. Here the message is one of mutual respect between rulers, and the content and presentation must be calibrated very precisely so that the other party doesn't feel either slighted or embarrassed by the gift.

In Western culture, the story of the three kings who came to present the infant Jesus with their gifts of gold, frankincense, and myrrh is powerful because it represents a reversal of the usual procedure by which the rich and powerful expect tribute from the poor or defeated. Artists have generally shown these gifts in the sort of chest a monarch might expect to receive.

The gifts of the Magi are one justification for the orgy of buying and giving that surrounds the Christmas holiday. Modern Christmas and modern packaging grew up at the same time, propelled by many of the same forces. Christmas-gift packaging for such items as biscuits and whiskey came early; people are generally willing to pay more for the packaging of something they intend as a gift than for an item they buy for themselves. The wrapping papers, ribbons, and adhesives that we use to disguise the packages in which gifts come are products of the same technologies as the packages themselves. Yet the gesture of covering packaging that talks about the product with wrappings that are part of the larger festival is important. The ritual of gift giving may be only a thinly disguised excuse for consuming even more intensely. But by wrapping the gifts, people identify themselves not with products but with families and friends, and the gift giving does become a ritual of tribute and cohesion.

In Japan, such rituals remain alive, and packaging that requires handcraft and uses natural materials such as leaves or paper-thin slices of wood can be found in department stores. There, a high proportion of consumption is channeled into gift giving, especially during the two official gift seasons. This manifests itself in very beautiful packages, including impeccably composed, well-made presentation boxes even for seemingly mundane, staple products such as noodles. At Zen temples and Shinto shrines, one often sees offerings on the altars that are right off the retailer's shelf. This concern with proper packaging leads to what seems to outsiders to be extreme overpackaging, something that certainly contributes to the solid-waste problems of Tokyo and other large Japanese cities. The obvious solution is to reduce packaging at least to American or European levels, something that is starting to happen. But because packaging is part of culture, changing it may have unexpected impact. If packages become less suitable for giving to others, will people buy them for themselves, or simply not buy at all? Gift giving is an expression of politeness and an instrument of cohesion, both of which are part of Japan's image of itself. Some in Japan are promoting new traditions of "gifting" between individuals, rather than in the family and institutional contexts in which most gift giving is done now.

On the subject of tribute and protopackages, it's worth mentioning another ancient story: that of the great wooden horse presented by the Greeks to the city of Troy, as recounted in the *Iliad*. The Trojans had no idea that there were Greek soldiers hidden inside when they brought the horse within the walls of the city. That was, of course, what the Greeks were counting on. It was a false tribute, a statement of respect made in bad faith. It dramatizes a fundamental lesson, and it is one that most shoppers probably have somewhere in their minds when they go to the store. We take packages into our homes and make them part of our intimate lives. Is the package really what it claims to be? Will it, like the Greeks' gift to Troy, contain any unpleasant surprises?

31

*   *   *

"Good things come in small packages" says the cliché. This statement is actually part of a more general truth, which is that precious things come in packages. Things that are very expensive tend to be purchased and used in very small quantities, and it is necessary to hold such commodities in a way that keeps them from being wasted. Moreover, there seems to have been a desire throughout history and across cultures to express the preciousness of the contents by making the container special as well.

The protopackages that are easiest for contemporary consumers to identify with are those that held cosmetics and perfume. Aromatic essences remain fairly expensive, and even when they are not, they are thought to be cheapened by being placed in large containers. Moreover, perfuming the body and painting the face suggest a kind of magic that we still hope will work. People use perfumes and cosmetics to transform themselves. Special containers embody the preciousness of scents and tints, while defining and expressing their power.

Containers for face paints and scents have been made from many precious materials, including gold, ivory, alabaster, jade, crystal, and porcelain. At many times and places, the containers for scents, paints, and unguents have been so finely worked and delicate that they approached the refinement of jewelry. Indeed, they did contain body adornments that were akin to jewelry. (This identification reached its logical conclusion in nineteenth-century France with the introduction of necklaces made of hollowed-out pearls, intended to hold perfume.)

Perfume and cosmetic containers — especially those made of glass — provide the most direct connection between the ancient world and the invention of modern packaging in Europe and the United States during the last two centuries. This is not simply because the making of perfume and its containers was a skill associated with the courts of nobles and monarchs in seventeenth- and eighteenth-century Europe, just as it was in Egypt thousands of years before. It is also because the crafts of glassmaking and perfumery can be traced from Egypt and elsewhere in the eastern Mediterranean through Rome and Byzantium and the Arab Empire and to Europe following the Crusades. From medieval Venice, glassmaking diffused through Europe, and in eighteenth- and nineteenth-century France, perfume and cosmetic containers became a strong influence on the creation of packaging as we know it.

Tracing this continuity is possible because although glass-making is a very ancient technology, it has also been a very static one. Glass objects were in use in Egypt and Syria before 3000 B.C. They were made by shaping silica paste around a metal rod. A bottle was made either by winding molten glass around this core or by dipping the core in molten glass. Once the glass cooled, the paste core was dug out. Such a technology does not encourage the extremely narrow-necked bottles we associate with perfume. Nevertheless, distinctively shaped bottles were created, and for some periods it is possible to identify scents, pomades, and face paint with containers of particular shapes: spherical, flat, tall and thin. Such a standardization of shape was hardly branding as we know it today, but it did provide a tangible identity for something that was bought and used in a small quantity and probably helped people to avoid confusing substances that looked alike.

The Egyptians also made unique bottles, often in human or vegetable shapes. They used many different colors of glass and worked in various textures, even with basket weaves and stripes.

Throughout most of history, innovations in glassmaking have involved aesthetics rather than production. The goal of the glassmaker has been to produce beautiful color, brilliancy, translucency, unexpected texture, or arresting form, not simply to produce. The crucial productive innovation of the entire period between the invention of glassmaking, sometime before 3000 B.C., and the patenting of the automatic bottle-making machine, in 1901, came in about 100 B.C. in Sidon, in Phoenicia. This was the invention of the blowpipe, which eliminated the laborious sculpting and digging out of the paste core.

The blown-glass process undoubtedly made glass cheaper, as the quantity of surviving Roman glass would suggest. A lot of Roman glass has lines that appear clean and modern, perhaps a result of larger-scale production. Many perfume and cosmetic bottles that survive have trademarks, which suggest that the bottle and its contents were sold as a single unit rather than separately. While perfume bottles stamped with names appeared as early as 1500 B.C., these were probably the names of the people for whom they were made. The Roman bottles, by contrast, were blown into molds and bear uniform marks, so that it is possible to conclude that some form of branding was taking place.

That doesn't necessarily mean that particular merchants or perfumers sought to identify themselves with a particular bottle design. As late as the mid-nineteenth century, American perfumers gave their buyers the choice of both the kind of scent and the bottle in which it was sold.

Some Roman bottles were obviously designed as gifts for wives or sweethearts and bear images of hearts or wedding rings and the word "Amor." No doubt, such valuable vessels had probably been presented as gifts all the way back to Egyptian times, but these pieces insist that the recipient remember the act of generosity in an altogether more modern way.

(Incidentally, the Romans produced one brand-name product we know of today, the Fortis oil-burning lamp. It was produced in Lombardy up until the second century A.D., after which it was supplanted by apparently cheaper imitations from the provinces.)

The so-called Dark Ages that followed the fall of the Roman Empire in western Europe were not a great time for perfume or the bottles that contained it. But elsewhere glassmaking continued, and indeed, the glassworks at Sidon where the blowpipe was invented continued in operation until A.D. 1200. Thus the tradition that had inspired Roman glass was still very much alive at the end of the eleventh century, when the Crusades began. Some of those who went to fight in the Holy Land brought back bottles of perfume, a precious commodity that was easy to carry. A passage in the *Song of Roland* concerning a chest full of exotic perfumes from the East describes their scents as sweet, but lavishes far greater detail on the "multicolored wombs" of the silver- and jewel-stoppered bottles of perfume and the fine pottery of the vessels containing face paints.

Of course, scents have always been difficult to describe, which is one reason that the vessels that contain them are so important.

This contact during the Crusades gave rise to the Venetian glass industry in the thirteenth century. Techniques for making stained-glass windows were known throughout Europe, but blown-glass vessels were a Venetian monopoly, and Venetians ruthlessly protected their city's status as Silicon Lagoon. Indeed, the Venetian Republic guarded glassmaking secrets more assiduously than modern states guard nuclear secrets. The glass industry was confined to the island of Murano, and the glassworkers were treated as a privileged class. But they were guarded, and if they tried to leave, they could be killed. For centuries, the secrets of Venetian glassmaking passed from father to son. The glass was shipped throughout the world, but the skilled craftsmen remained on Murano.

Inevitably, over time, some of the glassmakers and their secrets escaped from Murano. The path of diffusion was among wealthy and powerful people and the royal courts. The Medici of Florence were patrons of a glassmaker who introduced his techniques to the French royal court when Catherine de Médicis became queen. At the beginning of the sixteenth century, Queen Isabella of Spain had a well-stocked dressing table, filled with beautiful, varied bottles and elaborate pots. Later in the sixteenth century, Rouen drew many artists from Murano, where they found they had a greater measure of freedom, though less prestige than before. Still, the courtier class in France began to use large quantities of exquisitely bottled scents and cosmetics. England didn't produce perfume or glass bottles, but by the end of the century Queen Elizabeth I and her court were importing vast amounts of perfume from France and Italy and by some accounts were dousing everything in it, including dogs.

In France particularly the glassmaking craft was driven by that of the perfumer and the enthusiasm of the court. Louis XIV's Versailles was known as the perfumed court, and the enormous visual elaboration of the period was evident on the containers of perfume, makeup, and the newly popular *eau de Cologne.*

It's likely that many of these bottles for monarchs and nobles were one-of-a-kind creations, made to flatter the only people who could afford such luxuries. This is courtly craftsmanship, not packaging. Yet the desire to have an outward sign of one's ability to spend money on a luxury is an important part of modern packaging. This is true not only of perfume but of liquor, candy, and a host of luxury items. But before all consumers could pretend that they were living like kings, kings had to live like kings. Highly elaborate containers were part of the royal lifestyle.

At this point, we have to backtrack a moment for an important technological innovation — the patent of the coal-burning glass furnace in England in 1611. Previously, all glass had been made on wood fires, which produced glassware that was delicate rather than durable. The heat generated from coal was more intense and continuous, encouraging more rapid production of dark bottles, which were believed by wine makers to be superior. Wood-fired glass could not survive the pressures created by such effervescent beverages as beer and champagne. In England, this innovation spurred the exploitation of the Welsh coal fields, but it was in France that the coal-burning glass furnace had the greatest impact, as wine producers began bottling their products on a new and larger scale. The court-related glass craftsmen embraced this innovation, along with a related one, the development of brilliant lead crystal, and continued to produce beautiful, luxurious flacons. Meanwhile, the bottle was being transformed from a luxury good to a convenience to assist in the marketing of other products.

It would, however, be wrong to understand the story simply as a high-energy, high-production, wide-market technology supplanting a wood-fired, labor-intensive, narrowly based craft. The container as an expression of luxury is with us still, and indeed, the French are still an extremely powerful force in that market. Now, however, luxury is intended to be accessible to a much larger segment of the population, and the symbolic qualities of the container may, in fact, be even more important. The intrinsic value of the container has gone down; nobody sells perfume in bottles with amethyst stoppers anymore. That means that the design must be exceptionally creative to convey a sense that having and using the bottle is a real privilege.

When the French Revolution came, it did not destroy the Louvre, but rather turned it into a public museum. This transformation made the Louvre more central to French identity and experience than it had ever been as a royal palace. Similarly, the Revolution did not reject perfume, but rather attempted to democratize it. There was even a perfume called Guillotine, though this proved not to be a product people wanted to splash on their necks. During the early nineteenth century, perfume became the subject of considerable scientific research and emerged as an industry in its own right. The rise of branded perfumes had its first impact on the printing industry, as makers sought to create elaborate, colorful, and memorable labels to adorn a range of standard bottles. But by the mid-nineteenth century, just as modern packaging was beginning to emerge, the perfumers began to offer their products in bottles of striking, distinctive shapes. Often they were sold in wooden boxes, lined with silk or velvet and decorated with enamel. They were unquestionably a luxury product, but those who purchased them were no longer patrons, merely consumers.

The evolution of containers for such tangible contents as beer, wine, eye shadow, or perfume leads most directly to the commercial packages we know today. But they are really only one part of the story.

One of the key roles of all packaging is to make those who see and use the package believe in the efficacy of the contents. A contemporary package might try to convince people that a cleanser will really clean or that a cereal will really promote good health. But somewhere in the background lurks religion. Many, though admittedly not all, religions organize their worship around different sorts of containers of sanctity. These have come in all sorts of shapes and sizes, from the small vessels in which many Chinese still make snack-sized sacrifices of food to the ancestor spirits, to the pyramids of Egypt (which long ago failed the package's basic task of protecting its contents).

The container around which most synagogues are designed contains holy scriptures, which most members of the congregation know intimately. It recalls the original ark of the covenant, the original repository of God's written word. The way in which the contents of the sacred works exert their power — through the lives of the people who believe in and act on the sacred words — is not at all mysterious. Moreover, the cabinets used in most synagogues are usually of an appropriate size to contain the scrolls and, although they are usually well made and highly ornamented, they do not attempt to upstage what is inside. The container is thus a symbol of shared respect for the contents, which are well known to the worshipers.

Such a rational view of the relationship between a ritual container and its contents is, however, more the exception than the rule. Scriptures suggest that the original ark — whose design had been dictated by God — was carried into battle and was a source of enormous power. The popular film *Raiders of the Lost Ark* represented this container as having the potency of a neutron bomb. The memory of this supernatural power is embodied in the arks found in synagogues today. But the impulse to worship this object is undermined by one of the commandments found within: God's prohibition of idols and graven images. The Bible shows that God was right to worry about the human impulse to exalt the shiny object over the idea. Our tendency to transgress this commandment is something packagers count on.

Most religions, cursed with gods less articulate than that of the Hebrews, require more of their religious containers. Such vessels must impress two distinct constituencies, both God and man.

Most religions begin in the idea of offering sacrifices to placate powerful forces that control human survival. Sometimes, despite such sacrifices, the rains don't come and crops wither, or floods come and towns are washed away. People are always anxious about whether the sacrifices they made are good enough to do the job. Inevitably, people begin to deal with the deity as they do with a rich uncle. They embellish the gift by surrounding it with pomp and ritual and creating special richly decorated vessels made of precious materials. Such valuable vessels have sometimes been buried along with a powerful leader, but most often they have been reused. With rare exceptions, communities cannot afford to throw such wealth away.

For the human constituency, the worshipers, the lavishness with which the offering is presented reassures them that they are doing enough; at the same time it impresses them about the power of the deity to warrant so rich a display. In only a few generations, the caskets, cups, cabinets, urns, and other ritual vessels can accumulate into a show of wealth greater than that attainable by any individual. Such a treasury can become a center of power in itself.

For a religion such as Christianity, which keeps the idea of sacrifice at its center even as it has eliminated literal blood sacrifice, the container inevitably becomes more important. In Roman Catholic and Orthodox Christianity, the bread used in the ritual does not merely stand in for flesh in a sacrificial ritual, but is believed to be the true body of Christ, God himself. It is at once an offering *to* God, an offering *of* God, and, visibly, a piece of bread. The container that holds this transformed bread must somehow express that this is more than bread, and it ought to be visible all the way from the back of the church. The vessel used for display of the holy bread is called a monstrance, and in the treasuries of the great churches of Europe, you can see these made from gold and covered in precious stones. While the bread itself is believed not to be symbolic, but truly divine, the container, with its show of worldly wealth, is meant to be symbolic of the bread's supernatural power. The monstrance is not meant to be worshiped; its job is to call attention to the invisible presence of Christ within the bread.

In the interest of selling a product, rather than partaking of God, modern packaging makes some similar gestures. Like the precious stones of a monstrance, the red, yellow, and blue balloons on a Wonder bread wrapper were designed in order to catch the eye from a considerable distance and command recognition. The white background of the wrapper is intended as an image of physical purity, while the gold of a monstrance conveys purity of a more precious and exalted kind. And the combination of the white wrapper, the colored balloons, and the bold lettering is intended to communicate a sense of animation. This childlike energy and enthusiasm draws attention to something that is not visible in the bread: the nutritional enhancement that is said to build strong bodies twelve ways.

In the Catholic Church, relics of the saints — which can be bits of bone, small pieces of fabric, chunks of wood, and other items of little visual impact — have, over the years, demanded charismatic containers. During the Middle Ages, remnants of saints inspired passionate devotion and inspired people to take long-distance pilgrimages to visit them. Reliquaries had to be designed to give pilgrims a sense that they had arrived. Besides, one of the chief qualifications of saints is that they can perform miracles, a power that does not end after death. Indeed, most miracles are posthumous, coming as the result of prayer, often in the presence of relics. Thus, the reliquary does not merely protect and give visual expression to a precious though inert object. Rather, the relic is an active ingredient, a tool that the devout can use to literally accomplish the impossible. Thus, when you see a medieval reliquary in the shape of an arm, that doesn't mean it contains a fragment of an arm. More likely, it is a relic of a bishop, and the reliquary represents a gesture characteristic of the office. Like many packages, the reliquary expresses not its ingredients but its power.

The most potent expression of the power of the relic is architectural. The possession of an important relic made Chartres Cathedral possible — and necessary. The cathedral is a lot more than a package. But it contains other containers — reliquaries and tombs — that rely on art and permanent materials to express the transcendent qualities of crumbling bits of organic matter. Designing a vessel to express the meaning and force of a substance for which those qualities are not visible is the essence of a packaging job.

One of the greatest of all reliquaries is the Cathedra Petri, the Throne of St. Peter, in the apse of St. Peter's Basilica in Rome. It was built during the 1660s by a small army of metalworkers, glassworkers, sculptors, and other craftsmen, according to a design by Gian Lorenzo Bernini. This remnant of a chair is housed in an object that looks like a floating throne, seemingly levitated by the gaze of St. Augustine and other doctors of the church. It also contains an unexpected window that admits a mysterious light. It doesn't look like a package; it's almost a building in itself.

This work, part of a campaign to reassert the divinely ordained power of the papacy, is a landmark of the baroque style, of which Bernini was an inventor and a master. The essence of baroque design is not the expression of eternity or repose, but of energy and change. Bernini's specialty was to freeze the crucial moment of transformation in a way that draws in the viewer's eye and body and engages the emotions and spurs the imagination forward to the probable result. His work is explosive. It expands in space with enormous force.

Package designers face similar challenges in endowing their creations with visual and emotional qualities that make them distinctive on the shelf. And they have, over the years, used many of Bernini's graphic tricks to animate their packages. They use diverging lines of force to indicate their product's power. They use rippling ribbons to give inert, boring shapes a sense of movement and to provide a place to put written messages. (This one has medieval antecedents, but the contemporary usage is closer to baroque.) They use incomplete geometric shapes, which they invite the eye to complete, thus increasing the perceived size of the package. They use ovals rather than circles when they want to convey a feeling of potential energy and surprise. Thus, although the letterforms of such post–World War II detergent boxes as Tide, Cheer, and All may have been modern in derivation, the underlying spirit is clearly baroque. And although more recent package designs tend to be a good deal less over-the-top than the designs Bernini provided for Counter Reformation popes, much of the spirit of energy and manipulation of the artists of the baroque endures on the supermarket shelf.

Some precursors to packaging have existed far longer than people have. These are biological forms that protect or propagate life. The egg may have come after the chicken, but it certainly came before the L'eggs pantyhose package (and Silly Putty came somewhere in between). Packagers speak of clamshell containers and of peanut flowables, the low-density plastic packing stuff that falls out of shipping cartons when you open them. Each of these bits of modern packaging technology derives some inspiration from natural phenomena. Whenever you consider the nature of packaging, the fruits and the nuts — and the birds and the bees — turn up eventually. We should resist the temptation to call these natural analogues, or anything else discussed in this chapter for that matter, the first packages. Still, by looking at biological forms, we can learn something about the nature of packaging.

The most obvious place to start is with things found in nature that seem to be already packaged, such as nuts in their shells, peas in their pods, fruits like oranges in their rinds. These biological adaptations perform some of the tasks of packaging, primarily protection. For example, the shell on the nut allows the seed within to remain dormant for a long time and for it to be carried far from the tree from which it came. Eggshells also have a protective function, but they are most often admired as an economical and beautiful form, well adapted to the physiology and behavior of the species that lay them. The egg is a paradigm of "good design." It's probably too attractive, however, to predators, such as ourselves.

It seems natural to think of a nut or an egg as a package, because neither is too hard to crack and both contain nourishment. Natural forms whose protection is more effective are less often viewed as precursors to packaging. For example, people seem always to have eaten oysters and clams when they were available. But from the dawn of time until now, the shell has been a nuisance. Because it serves the mollusk better than it serves people, we don't see the shell as a package. Clamshell packages, the hinged, two-part, closable boxes that hold sandwiches in many fast-food restaurants, echo the bivalve's configuration of two shells with a hinge, though it's unlikely that most people think of clams when they see the package. (People are more likely to worry about litter or damage to the ozone layer.)

An orange is more like a package than is an oyster because it invites participation. When a creature eats the fruit, it carries the seeds and disperses them, thus contributing to the propagation of future orange trees. When the oyster shell is cracked and the animal inside is consumed, the reproduction of the species is not enhanced.

But when the orange peel is split and the seeds within are dispersed, the reproductive cycle of the orange continues. Protection is a role of packaging, but it is not the only one. Packages are made to be opened.

One other important thing about oranges: they are orange. The use of a single word to name both the fruit and its color demonstrates that the fruit has a strong visual identity. Like a good package, it communicates, and it happens to have an extremely eye-catching color. One indication that the orange has virtually become a package in the modern sense is the practice by some fruit shippers of dyeing oranges whose color is deemed to be insufficiently orange. Thus nature's orange is modified to make it the package the marketplace expects.

In the biological world, flowers are the great communicators. We don't intuitively associate packages with flowers, but it's worth considering. That's not because they're pretty, with colors that attract the human eye. The audience for flowers is insects, and to a far lesser extent birds, that are drawn to their nectar and, in the course of getting some, cross-pollinate the flowers.

Flowers and their pollinators have evolved together over the last 225 million years. The first flowerlike buds were very tiny, only about five millimeters across. Much larger flowers began to appear about 140 million years ago, at around the same time as the emergence of birds. True flowering plants, with colorful variegated flowers, appeared about 100 million years ago, along with bees. Insects have a very different kind of eye than vertebrates do and are sensitive to a somewhat different spectrum. Bees cannot see red, but they can see ultraviolet colors that are invisible to humans, and it is not surprising that many flowers have patterns in the portion of the spectrum that can be seen by bees but not by people.

This coevolution has resulted in many subtle relationships between the flowers and the pollinators. As with packaging, mere visibility is not enough. Many flowers have evolved so that they can only be pollinated by insects of a certain species and time their blooms and release of scent to assure that inappropriate insects won't be attracted. There is obviously an advantage in making insects loyal to a particular species, so that flowers will not keep getting brushed with pollen that won't do their plants any good. There has also been a precise calibration of the amount of nectar flowers provide for the insects. If there is too much from each bloom, the flowers will not be thoroughly cross-pollinated. Too little, and the insect does not have the inclination or the energy to go on. The travels of a bee through a meadow full of flowers has been studied almost as closely as the path of a shopper through a supermarket, and there are some similarities. Both bees and consumers face a race between stimulation and exhaustion. The shopper's eye spends only one-sixth of a second on a package, while the bee, whose commitment is stronger, spends far longer on each flower. Still, they can stop at hundreds in an hour.

Supermarket shoppers are a good deal more rational than bugs drawn by instinct to a particular kind of blossom. Insects exhibit a level of brand loyalty of which marketers can only dream. But package designers use many of the same devices of color, pattern, and shape to short-circuit consumers' conscious minds and induce engagement with the product through its package. Their goal is not so much the safeguarding of the particular contents of the package but promoting widespread, repeated use of the product. Their goal is not to make the package jump off the shelf, but to attract the sort of people who are likely to be interested in it and perhaps make it part of their lives.

And just as evolution has slowly brought changes in the forms of blossoms and in the bodies of the creatures that pollinate them, packages and consumers are engaged in a process of rapid social coevolution. The way people live changes packages, and sometimes packages change the way people live.

The natural phenomena that are most relevant to packaging — fruits, nuts, pods, flowers — are reproductive, rather than merely protective like the oyster shell. Packages are disposable, a means to an end: selling the product and keeping the productive system going. And although it is misleading to push evolutionary analogies too far, a supermarket is an intensely competitive place where any small advantage can ultimately yield large results.

If, as Calvin Coolidge said, "Advertising ministers to the spiritual side of trade," packaging meets its physical needs. The literature of package design is suffused with eroticism. Packages reveal and conceal. They beg to be unwrapped. And then a moment after consummation, they lose their magic as they turn to trash. Then very soon the cycle begins again.

Some packages are, of course, clearly sexual. One study indicated phallic lipsticks sell better than less aggressive ones — except that those that are unambiguously so don't sell well at all. Aerosol cleaners and pump sprays project power in a way that's clear, though not blatant, and some other cleaning products come in containers whose broad-shouldered shape connotes strength.

Far more products, including those aimed at men and women alike, are feminine in form. The classic six-and-a-half-ounce, thick glass Coca-Cola bottle has often been said to have an affinity with prehistoric carved female figures that are assumed to have been associated with fertility rites. It is always described as the hobble-skirted bottle, which clearly describes both the form of the female and her subjugation.

A cinch-waisted shape is easy for the hand to grasp, which helps explain its popularity, but a female shape also tends to be more reassuring than a male one.

Those who design packages want people to feel comfortable with the product, and packages that suggest the female form speak directly to infantile emotions. They are less often sex objects than mothers. They suggest the first fulfiller of needs, the original source of satisfaction.

Perhaps they even evoke prenatal memories. After all, Mom is the package we came in.

3 *Trusting the package*

The first modern packaging — named products in distinctive containers with labels — seems to have arisen in London around the turn of the seventeenth century. The packages contained medicines — elixirs, salves, and ointments meant to treat specific complaints or cure a wide variety of illnesses. Sometimes they worked on horses or cows as well.

The technology of the packages was primitive by contemporary standards. The bottles were handblown, without even the use of molds to shape them. They were wrapped in labels made from paper that was handmade sheet by sheet and printed on a handpress. Some medicines were packaged in ceramic pots that were also laboriously made. Crown caps and screw tops were still two centuries in the future. There is every reason to believe that it was a great deal easier and cheaper to mix the chemical compounds contained in these packages than it was to make the package itself.

The brands — including Stoughton's Drops; Singleton's eye ointment; Lockyer's, Hooper's, and Anderson's pills; British Oyl; Turlington's Original Balsam; and Daffy's Elixir — became familiar names in the English-speaking world for close to two hundred years. These were do-it-yourself remedies created at a time when medicine itself was based on the theory of body humors, which was not a particularly helpful diagnostic or prescriptive tool. Herbal medicine and other traditional remedies discovered by trial and error were on the whole more effective than those that depended on the medical science of the day.

However, these proprietary medicines were not, for the most part, based on traditional herbal remedies. They represented a far more modern, though often less effective, form of magic. Their verbose wrappers, which were often not labels but poster-sized broadsides folded around the bottles, made a strong case for the scientific respectability of these cures. Physicians' endorsements, along with testimony by miraculously cured patients, were mainstays of medicine promotion. Anderson's pills were introduced in 1630 by a Scot who claimed to have been personal physician to Charles I, thus beginning a long tradition of endorsements by monarchs and celebrities. The actual contents of these preparations changed over time. The one thing that remained constant was the shape of the bottle and the name on it. The Turlington's bottle was pear shaped with high shoulders, and the Godfrey's Cordial vial was shaped as a truncated cone with steep-pitched sides. Purchasers were cautioned to beware of imitations.

Just why these remedies arose at the time they did is not really clear. One reason was probably the introduction of the use of coal in bottle making discussed in the previous chapter. This increased bottle production and brought the cost down. Similarly, printing was becoming more efficient, so medicine producers were able to afford to wrap their bottles in discursive wrappers. Literacy, though it was hardly universal, was on the increase, so it was likely that potential users could read at least some of the pseudoscientific verbiage on the labels.

Increasing urbanization tended to break family ties and leave people on their own without access to traditional cures. The colonization of the Caribbean and North America sent people to remote locations, farther removed not merely from the expertise but from the plants on which traditional medicine was based. These remedies were shipped to the colonies and were apparently very reassuring.

The Total Package

The development of patent medicines coincided with the rise of Puritanism and the English civil war, and there was probably a relationship. In contrast to a hierarchical church with centralized authority and teaching based on ritual, image, and oral interpretation of the faith, Puritanism trusted individuals. They could read and learn from the Bible on their own, the Puritans believed, and the congregation was defined by a gathering of faithful individuals, rather than as a unit of a larger bureaucracy. Although Puritans were not the most liberal of thinkers, their belief in individual inquiry helped spur literacy and the advances of seventeenth-century science. Today, it may not seem obvious that patent medicines were part of the same movement that produced Isaac Newton. But they did represent a move away from oral traditions, and though they may not have been new discoveries, they proclaimed their allegiance to the world of science.

It's also worth noting that the concept of individual sovereignty set the stage for a new kind of human — the consumer. In a faith where people are, at least theoretically, free to make their own decisions, it is not enough that they should be told what is right. They must be convinced. Puritan communities were coercive. But Puritan doctrines placed great emphasis on a personal relationship with a distant God — not mediated by an army of middlemen. In the same spirit, manufacturers of packaged products seek to create a relationship with the ultimate user in which retailers and wholesalers play a very small role. Consumers trust distant, unseen manufacturers more than they trust the neighbors who dispense the product. (Puritanism also contributed to the growth of the consumer society, through its belief that material well-being might be a signal of God's favor.)

The patent medicines may also have been a way around community disapproval of liquor. Most patent medicines contained large quantities of alcohol, which was perhaps the chief ingredient that made people feel better. And over the years, most people who used patent medicines were aware of this intoxicating ingredient, though they pretended not to be. Daffy's Elixir was created by Thomas Daffy, a provincial clergyman, for children. But so many adults took Daffy's, augmented with gin, that by the middle of the nineteenth century in England, "daffy" was a slangy euphemism for gin. The connotation was that the person who only had a bit of Daffy now and then lived in an alcoholic haze, but never admitted drinking. In mid-nineteenth-century America, makers of highly alcoholic medicinal bitters sold their products in decorative bottles. But very few of these survive, very likely because those purchasing the product were aware of the alcohol and did not wish to save or display trophies of their own hypocrisy. Many stores sold bitters either by the bottle or the drink.

These British preparations were widely available in the North American colonies before the Revolution, and indeed, the only evidence of a homemade product is an advertisement in Benjamin Franklin's *Pennsylvania Gazette* for "the Widow Read's ointment for the ITCH." Since Mrs. Read was Franklin's mother-in-law, it's possible that Franklin himself was trying to get into the lucrative remedies business. Another renowned colonial printer, Peter Zenger of New York, published what is believed to be the first American medical pamphlet. It was essentially an encomium to an English product, Dr. Bateman's Pectoral Drops, which, the pamphlet said, cured gout, rheumatism, jaundice, stone, asthma, colds, rickets, and melancholy.

Actually Americans were in the nostrum business, but they used the British bottles and labels. In 1754 a Williamsburg druggist ordered more than 1,700 Stoughton's bottles and a large quantity of the labels from a supplier in England. There is no evidence that he also imported the contents. A few years later, the maker of Turlington's complained publicly that "New York scoundrels" were purchasing empty bottles of his medicine and reselling them filled with "a base and vile composition of their own." Proof of the truth of his grievance came during the Revolution, when all trade was cut off between the warring countries, but the sale of English medicines went on without interruption. And after the war was over, when trade was restored, the English brands dominated the American market, but they were mostly American-made contents in the English bottles. The packaging did not merely sell the product. In the minds of buyers, the packaging was the product. Stoughton's even entered the American language. During the mid-nineteenth century, someone who "sat there like a Stoughton's bottle" was inert and ineffectual.

Americans were not the only ones to try to profit from copying the overall look of a package, which lawyers call trade dress. In 1760 the manufacturers of Singleton's eye ointment, by that time a 163-year-old brand, brought suit in London against several parties who had counterfeited the printed directions and copied the distinctive, pedestaled ceramic container. The principle of protecting trade dress had been established even earlier, in 1623, when James I granted Roger Jones and Andrew Palmer a patent for hard and soft soap that included exclusive right to mark on goods or containers a rose and crown. This patent went far beyond any later package and brand protection by allowing Jones and Palmer to enter, with assistance of constables, anyplace where they had reason to suspect that their product was being counterfeited. Nevertheless, counterfeiting and imitating of packages and trademarks has continued to this day, punished less often than it is tolerated. Patent medicines remain a major target for imitators.

The first successful American patent medicine — Lee's Bilious pills — was developed by Dr. Samuel Lee of Windham, Connecticut, sometime before 1796. By 1857 there were 1,500 different patent medicines sold in the United States. These weren't true patents, granted by the Patent Office to innovative inventions. Before 1870, when the Library of Congress began trademark registration, names, logotypes, and distinctive packages enjoyed no specific protection, though courts had recognized them as a form of property.

These remedies ranged from the alcoholic cordials to ketchup (which, as Dr. Miles' Compound Extract of Tomatio, was hailed as a health miracle in the 1830s) to Smith Brothers cough drops, featuring the likenesses of the bearded siblings on the box. At this time, Yankee peddlers who traveled the countryside with carts full of tin pots, cloth, spices, and other treats and necessities were an important part of the American commercial scene. Patent medicines were an extremely attractive product for such merchants because they were light in weight, small in bulk, and high in value.

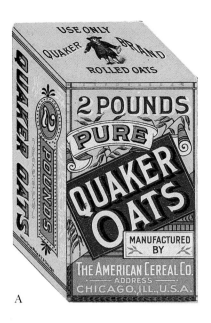

A

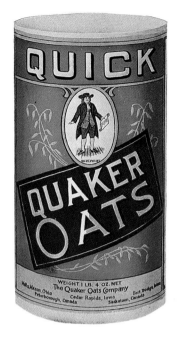

B

C

D

Quaker Oats, the first packaged cereal (A), appeared in 1884, only a few years after the invention of the folding box, and later adopted its distinctive cylinder box (B). Packaging is crucial to cereal, to express freshness (C) and provide mini-billboards for the product (D).

2

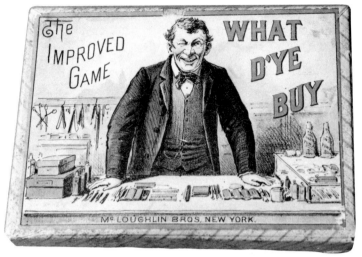

A

In the late nineteenth century, merchants weren't always trusted (A), and as this 1887 trademark (B) demonstrates, there were important things to worry about. The dog and cat that can't get at the meat in this trade card (C) are stand-ins for less endearing animals.

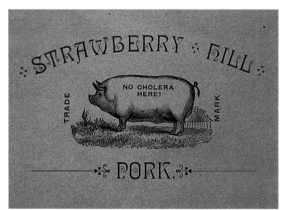

B

C

3

The friendly grocer stood between the shopper and the product in 1912 (A). By 1933, he was only around to help (B).

This Oval Label Protects You and Your Dealer

ARMOUR AND COMPANY

A

B

A

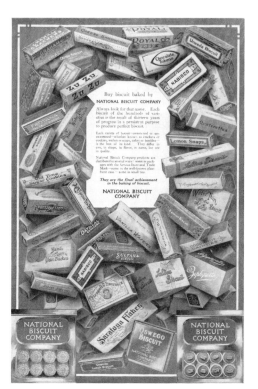

B

The raincoated Uneeda boy (A), introduced in 1899, embodied the product's protective packaging. This sales pitch was rapidly expanded to other crackers and cookies, as this 1912 ad shows (B). By 1950 the In-Er Seal symbol had become Nabisco's corporate logo and was used on clear cellophane packages (C); it was later modified by Raymond Loewy (D). Current hot cookies include guilt-free SnackWell's (E), whose green packages depart from the warm colors identified with the product category. A private label cookie (F) uses clean graphics and a lack of sensuality to establish a premium image.

C

D

E

F

6

A

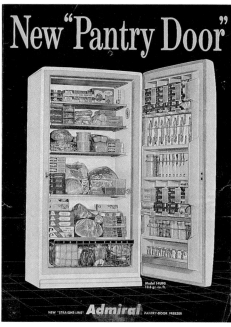

B

Frozen foods were invented in the 1920s but were still a novelty in 1946 (A). The frozen TV dinner became a mid-1950s symbol of modernity, and stand-alone freezers were treasure-houses of edibles (B). A contemporary children's dinner (C) shows how markets have become specialized and fragmented.

C

Why do you suppose the manufacturer put a box of Tide in your new automatic?

"I recommend Tide for every automatic," says John E. Ryan of Hempstead, Long Island, N.Y. "Folks ask so many questions about what's best for their machines! And I've discovered that to start people off with Tide gives them the kind of clean washes that makes them bless the day they bought their automatic."

...so your automatic will give you the cleanest clothes possible!

A

B

Tide detergent was popularized along with the automatic washer, and its box was designed to evoke power (A). Pioneer marketing psychologist Louis Cheskin, who helped design the box, liked to demonstrate its recognizability even when the image was scrambled (B). Today the product comes in shorter, concentrated boxes (C), plastic bottles (D), and bags that promise 80 percent less packaging (E).

C

D

E

8

McDonald's golden arches marked a distinctive package on the 1950s and 1960s commercial strip (A). Even when the design is toned down to blend into a neighborhood (B), it's still unmistakable. Inside, the clerks are part of a foolproof marketing and service system (C). The french fries box is designed to always appear to be overflowing (D).

A

B

C

D

Besides that, they were very attractive to people in remote locations far from doctors, and often far from friendly advice. The medicines were literally wrapped in reassurance, the small-type instructions on how to deal with a wide range of terrifying ailments.

Still, a lot of the people who bought these medicines lived in towns and cities where they were able to see doctors. Some of the patent-medicine makers implied, none too subtly, that doctors shunned their product because it worked too well and would be bad for business. One advertisement featured a doctor, "more honest than the rest," who explained that he used to prescribe a patent medicine, but stopped because patients didn't come back. "If I had kept on, they would have been taking BAD-EM SALZ and getting well without my assistance." Although they rarely make their attacks so bluntly, makers of packaged goods often seek to encourage mistrust of people close to the buyer — the corner grocer, the store salesclerk, or parents and family. The primary relationship is between the manufacturer and the buyer, through the medium of the package. As Bandreth's pills, an early-nineteenth-century American remedy, sold mostly by peddlers claimed:

> Whether it be a cold or a cough, whether it be asthma or consumption . . . whether it be typhus fever or ague, or bilious fever, cramps, whooping cough or measles; whether it be scarlet fever or small pox, Bandreth's Pills will surely do more than all of the medicines of the Drug Stores for the restoration of your health.

Sometimes, despite their reassuring words and familiar containers, patent medicines were deadly. One of the best packaged and most popular of the early American preparations was also one of the worst. In 1822, William Swaim of Philadelphia commissioned Alexander Anderson, a leading woodblock engraver, to make a striking rendition of Hercules slaying Hydra to serve as the symbol for his product, Swaim's Panacea. Previous patent-medicine makers on both sides of the Atlantic had generally refrained from claiming to be a panacea — a cure for everything. Often their lists of what they could cure were so long and unlikely that they seemed to be cure-alls, but unlike Swaim, they stopped short. Like Hercules on its label, the liquid in the squat, six-sided bottle claimed to be more than the equal of anything it went up against. Following one astounding cure involving a Nancy Linton of Chester County, Pennsylvania, who attributed her deliverance from death's door to a dose of the panacea, Swaim published a book detailing the cure and incorporating hundreds of professional endorsements. It is illustrated with a terrifying engraving of Mrs. Linton, sitting calmly in her chair with a bottle of Swaim's by her side. But the legs that emerge from her dressing gown are those of a skeleton, and it is easy to confuse her legs with those of the claw-footed chair in which she sits. Perhaps it was meant to suggest that the panacea cured from the top down, or to hint that death is always near, but it's difficult today to imagine that anyone who saw that image would be moved to run out and buy a bottle of Swaim's.

Those who did purchase Swaim's were not doing their health any good. It contained sarsaparilla, a common ingredient in patent medicines, and wintergreen, the first commercial use of this fresh, invigorating flavoring. Unfortunately, it also contained corrosive sublimate of mercury, a highly poisonous substance. A few doctors protested against this product and other products that were so highly alcoholic or narcotic that patients were able to believe they were getting well even as they were becoming sicker. But regulation of such medicines did not arrive until 1906.

One thing that medicine makers understood early was that it is helpful for a product to have a story. The entrepreneur responsible for Anderson's pills not only claimed a royal connection, but also spread a story of learning of the remedy on a trip to Venice. Things that were wild and alien were highly attractive. During the eighteenth century in England, bears, which were hardly common in Great Britain, appeared as commercial icons, promoting several different makers' bear's-grease pomades. These were packed in ceramic pots, which in one case proclaimed the pomade "well-known to be the only thing yet heard of for restoring decayed hair." There were also ceramic bears, made for what would today be known as point-of-sale displays.

When Americans started producing patent medicines, they tapped into the buyers' interest in the primitive and exotic by presenting their products as adapted from traditional, and secret, Indian remedies. The use of native North American medications reflected an assumption, made by botanists and physicians alike, that the cures for ailments are found close to their causes. Moreover, stories of Indians gave the medicines a strong personality, a sense of adventure and exoticism. This was particularly true for those who lived in the East, where encounters with Indians were a rarity.

For several decades, manufacturers and sellers of medicines had hawked their wares by incorporating their sales pitches into entertainments, a practice that prefigured commercial radio and television. At first these were solitary peddler-performers. One such was William Avery Rockefeller, who during the first half of the nineteenth century traveled through the newly settled areas of what is now the Midwest, performing as a hypnotist, ventriloquist, marksman, lecturer on herbs, and, not least, salesman for a number of patent medicines. (His son, John D. Rockefeller, didn't look like a ventriloquist's son, but he did use patent medicines all his life.)

Later, this sales approach became very elaborate with the emergence of full-scale medicine shows, consisting of whole troupes of entertainers. An advance man would plaster the targeted town with broadsides, and then the troupe would arrive as a band playing musical instruments. Their offerings included violinists, dancers, poetry readings, and disgusting displays of animal parasites they bought from slaughterhouses.

The trend toward traveling extravaganzas converged with that of high-personality products during the 1880s with the introduction of the product called Kickapoo Indian Sagwa. A typical Kickapoo troupe consisted of six Indians — it didn't matter what tribe, but they were really Indians — and six white men. The Indians danced, shouted, and performed in the ways that people expected Indians to perform. The white men did not so much perform as explain what the Indians were doing and reassure spectators that this encounter with savagery would not get out of control. The climax came when one of the Indians would give a lengthy and dramatic speech in his native tongue, which would be translated as a description of *sagwa*, the miraculous remedy of his people. Even while the performance was going on, members of the troupe went through the crowd selling bottles of it. At one time in the late 1880s, there were seventy-five Kickapoo troupes performing throughout the United States.

In contrast with this highly emotional approach, other traveling pitchmen sought an image of vaguely exotic sobriety. Uncommon religions that involved costumes were useful because they brought a sense of probity and trustworthiness to an undertaking that had a raffish and unsavory image. At the same time, the strangeness of the Shakers, for example, suggested that they might have some real secrets that were worth buying. Indeed, some Shakers did hit the road, selling herbal remedies the community had developed. Another common figure was the Quaker, with his broad-brimmed hat, plain good clothes, and use of the words "thou" and "thee." The medicine-selling Quaker was to be appropriated to lend his reputation for integrity to the product that pioneered the nationally marketed packaged-food industry.

Much of the late-nineteenth-century phenomenon that has been termed the packaging revolution consisted of marrying a host of technological innovations to ideas about defining and selling products invented in the often disreputable world of patent medicine. Indeed, some businessmen resisted the new marketing ideas of packaging and promotion precisely because they associated them with the slick, entertaining deceptions of nostrum sellers.

In fact, even those medicine sellers who were honest and who believed their product benefited mankind recognized that it was more important to sell an idea, a feeling, a reassuring package, than it was to sell the contents. There are countless chemical and botanical ingredients, flavors, and colorings that can be combined in any proportion. But these are not the only ingredients. A bottle that can be recognized on the shelf, a portrait of the inventor, and an engaging story are equally important. And packaging can have such a profound psychological effect, it can make the medicine work.

The packaging revolution got under way, primarily in the United States, during the last quarter of the nineteenth century. Certain brands had been established long before, of course. Pears in Britain and Colgate in the United States both started in the toiletry business around 1800. Keiller's Dundee marmalade began slightly earlier, and although its graphics have changed several times over the years, its pottery jar with transfer-printed label appears today as a living fossil from the age of Jane Austen. There were even some packages, largely for such high-value imported goods as chocolate from Baker's in America and Cadbury in Britain, that spread a brand and an image throughout the world. For many such precocious brands, the advantage gained in being the first to package their products has been sustained for close to two centuries.

But despite such anomalous pioneers, the idea of making a package known and familiar far and wide, investing it with a personality, and winning consumer loyalty would scarcely have been possible without many other innovations. Railroads made it possible to move bulky items rapidly and cheaply. Urbanization broke ties of family and community, making it necessary for people to trust strangers and to be susceptible to advertising in new, populist communications media. The move from subsistence living to wage-earning jobs allowed little time for people to make things for themselves and transformed necessities into consumables. The sheer productiveness of new machinery and production systems in turn made it necessary to consume items at a faster pace, merely to keep the machinery going.

Industrialization produced both a massive expansion of the number of things to possess and a broadening of the segment of society that could possess them. Today, most residents of developed countries live lives of casually promiscuous purchasing, and it is easy to forget that throughout most of history, most people in most places have lived with virtually no possessions. Even the well-to-do lived more sparely than the large numbers of objects on display in museums seem to suggest. The rich, well-crafted containers found in eighteenth-century American colonies for such products as tea, and even tobacco, hint that these were precious commodities. The objects don't suggest what is revealed on some tax surveys — that the majority of people living in the colonies that would become the United States had a pot, a few tools, and maybe one chair, but not two. For most people in Europe, life was not richer, and there was less food.

While industrialization did not spread its newly created wealth fairly or universally, it did so more broadly than before. And those who moved from subsistence agriculture to industry suddenly had cash. Their salary might not have been enough to allow them to spend it on luxuries, but cash confers the power to be irresponsible and buy what you want rather than what you need.

Industrialization centralized in factories the production of products that had once been handmade everywhere. While nearly all blacksmiths had made their own nails, machine-made nails were produced in a handful of factories and shipped widely. Such products had to be packed for transport. But how was the buyer of the nails to judge the product? He had previously depended on his familiarity with the skill of the blacksmith and now had to trust the merchant who was selling the nails. It was thus in the merchant's interest to determine that quality was consistent from one shipment of nails to the next. He often counted on the name of the manufacturer and of the product, which was literally branded on the wooden box in which the nails were shipped.

Industrialization brought with it not simply the movement of goods but of populations. It drew people from villages where they interacted with people they knew well to cities full of strangers. American cities experienced particularly explosive growth as they attracted massive European immigration, but even in Europe, small market towns transformed themselves into large industrial cities. Horror tales like that of Sweeney Todd, the London barber whose customers became meat pies, express the anxiety of newly minted city dwellers who didn't know what would become of them — and didn't know whether to trust what they bought. This urban population not only didn't know who was making the nails, but often didn't know whether to trust the merchant who was selling them either. In such a changing, unsettled environment, people naturally looked for familiar things they could rely on, and precisely the qualities that made people feel insecure increased their willingness to establish new loyalties.

It is probably accurate to say that until just a bit more than a century ago there were packages, but not packaging. Packages protected goods and made their shipment safer, and some even helped to sell the product. But the materials and technologies of the can, box, bottle, printing, and paper industries developed independently, the products of different histories and craft traditions responding to different needs. Only after the various technologies had reached a high degree of industrialization, and packages were sufficiently inexpensive to make and transport, did the power of packaging to sell products and control markets become apparent. By then it was evident that packaging, as opposed to bottle making or box making, was an important industry in its own right.

While pottery was the primary material for making containers in nearly all nonnomadic societies, and throughout most of history, its use in modern packaging has been relatively unimportant. Inevitably, the heaviest material has given way to those that are lighter in weight, are less expensive to produce, and can carry a lot of information. And that has, up to now, meant paper and paperboard (or cardboard). Folding boxes are by far the most used package in the world today, and various kinds of paper containers, most notably the brown paper bag, have played a major role in the selling and transporting of goods. Moreover, because paper is the easiest medium on which to print, paper has usually provided much of the information to be found on bottles, tinplate cans, wooden boxes, and other kinds of containers.

Paper was itself one of the first branded products. Watermarked paper was first produced in the fourteenth century, and both printers' marks and elaborately printed makers' labels for blank paper appeared during the sixteenth century. The first use of paper for wrapping was recorded in Germany shortly thereafter, in a form of recycling. Printers, who generally did not have books bound until they were bought, sold the pages from unpurchased books to merchants to wrap their products in.

We have seen that patent-medicine merchants, along with makers of some related products such as bear's grease, made use of paper labels both to sell their products and to convince purchasers that they had done the right thing. Paper wrappers were used in England from the mid-1600s for that very popular American-produced luxury good, tobacco. Still, some merchants would advertise high-value imported goods and add that the buyer should bring his own container. Personal tea safes were valuable possessions that held an expensive commodity. But they seemed like less of a luxury than each tea purchase packed in its own paper container.

That was because, until the end of the eighteenth century, paper really was a luxury, each piece made by hand. The first papermaking machine was invented in France in 1798, and it set off a chain reaction of innovation, culminating in the first machine to make paper in rolls, patented in England in 1807. Following the first American paper-roll operation in Delaware a decade later, the United States began to dominate the paper industry. The reason was the substitution of rags, which were scarce in the New World, by pulp from straw and wood, of which North America had plenty. The United States became the world's largest producer and consumer of paper by 1830, a distinction it still holds. This same period brought the first paperboard-making machine in England and the first board, made of straw, in the United States.

Also in 1798, in Germany, lithography was invented, making possible relatively low-cost multicolored designs. Together, machine-made paper and widespread lithography spurred a dissemination of vibrant imagery on a scale the world had never known. It was part of a social transformation that, by the end of the nineteenth century, enabled even the working class to lead cluttered lives. By 1834 there were seven hundred lithographers in London alone. (The entire United States did not match that figure until fifty years later.) It was no longer necessary to be rich for comely faces, pastoral vistas, or great moments from history, the Bible, or classical mythology to be framed on the wall of the hallway. And similar scenes were also appearing in less-expected places, first on match-boxes and then on the cans and cartons in the pantry.

It is striking that while many early printed labels consisted entirely of type, and perhaps a trademark, package imagery and color printing developed side by side. Unlike the movies or television, packaging did not pass through a significant phase of black-and-white images. While all-type black-on-white cartons persisted until the end of the century, packagers quickly understood the emotional impact of adding at least one color to the pictorial or symbolic elements of their packages. One of the earliest-known surviving American cans, made for tomatoes during the Civil War, sports a color label that, while it might not match the hyperreal imagery of contemporary package design, still has plenty of what food marketers call appetite appeal. Later it was discovered that vibrantly colored imagery provided an effective route around government regulations, when in 1888 American makers of butter substitutes were required to place their products in wrappers spelling out MARGARINE in bold, inch-high type. Most of them also included a picture of an attractive girl on the label directly above the word. "The joke of the matter," wrote *American Lithographer* at the time, "is 99 out of 100 people imagine Margarine is the name of the pretty young damsel."

Until about 1870, most packaged goods were luxuries, meant for gifts or personal indulgence, and color imagery was a large part of what made them worth more. Thus, although packaging was only a small part of the development of printing technology during the nineteenth century, it contributed to some very significant advances. For example, the offset-printing process was introduced in 1879 as a result of the desire to be able to print directly on tinplated biscuit containers.

Packaging also spurred the development of such paperlike materials as tinfoil, first used in France in about 1840 for wrapping individual candies. This material was superior to paper for preserving contents and stopping the diffusion of odors. In 1850 this led to a composite package for Cadbury chocolate that had a foil inner layer and an outer layer of stain-resistant paper. The concept of a package made of different layers, to maximize both the protection of the contents and the communicative power of the container, is still a mainstay of the food industry.

One of the most crucial innovations involving paper was also apparently one of the simplest — the paper bag. A rudimentary paper-bag-making machine was developed in Bethlehem, Pennsylvania, in 1852, and by 1860 a more efficient and practical machine for that purpose was being used in Philadelphia. But what ultimately spurred the use of the paper bag was the unavailability of cotton bags for northern millers during the Civil War. This cloth-sack shortage advanced paper bag technology, and soon Americans were exporting paper bags, and the machinery to make them, all over the world. In about 1870, an inventor named Luther Childs Crowell patented machinery to create the flat-bottomed paper bag. Like the Canaanite jar, the flat-bottomed bag became a universal container.

It is still an immensely useful part of life more than 120 years later. Flour still comes in paper bags, as do sugar, cookies, dog food, and other products. Moreover, shopping bags provide an important way in which retailers who do not primarily sell goods that come in packages can still create a physical identity that leaves the store and continues to sell on the street and in the home. But the greatest use of paper bags is probably as something to put packages in. They make it easier to carry things away.

Historically, paper bags preceded the popularization of the package. In fact, the paper bag, into which the grocer would scoop small quantities of sugar, rice, or other commodities from a big barrel, was what packages had an often difficult time displacing. Still, the introduction of paper bags represented an extremely important transitional step, because they were found to increase sales wherever they were used. It was no longer necessary to bring canisters to the grocery store to be refilled with staple commodities. Paper bags ensured that shoppers could carry home whatever they had bought. By reducing such barriers to purchasing, they acted, in effect, as a lubricant to allow the retailing machine to hum along at a continuous, predictable pace. Because the whole rest of the economy was being rationalized according to precisely these mechanical principles, the advantage of paper bags became immediately apparent.

"Nothing has had a greater influence in making possible the rapidity with which certain branches of the retail business are now conducted as compared with 10 years ago . . . than the cheap and rapid production of paper bags" wrote the economist David Ames Wells in 1889. By that time, American production of paper bags numbered millions per week. But it soon became clear that packaging represented an even more efficient accelerator for purchasing.

The medium by which packaging became a mass-market phenomenon was the paperboard folding box. Boxes are far less likely than bags to rupture during shipment, spilling their contents, which attracts vermin and lowers profits. Boxes are also far more effective than bags in keeping their contents from being crushed, which makes them attractive for products like crackers. Moreover, paperboard boxes are well suited for printing and stand straight and smart in store displays.

The folding box is so ubiquitous in contemporary life, and so straightforward in its concept, that it seems as if it must have always existed. The box itself is obviously an ancient idea, but the folding box is a direct result of continuous-process technologies. It is a creature of mechanized printing, filling, and transportation.

Box making was its own craft during the late eighteenth and early nineteenth centuries in both Europe and the United States. The box makers often made boxes both of wood and of paperboard, which was still a handmade material. Many of the board boxes were round or oval, a natural consequence of using a material that was easy to bend but difficult to crease and square off accurately. Their customers were jewelers, pill sellers, and others who sold low-bulk, high-value goods.

The finished boxes they made, called setup boxes, took up as much space when they were empty as when they were full. That limited the territory in which these boxes could be delivered, and the boxes were a nuisance, taking up a lot of space when they were waiting to be filled. The problem of managing empty boxes greatly limited their use, even after board began to be machine produced.

61

Setup boxes are still used for candies and many other gift items. But for everyday use, what was needed was a box that wouldn't become a box until it was time to fill it. The first folding box was produced in the United States during the 1850s by Bird & Company for carpet tacks. It was not a very satisfactory design because it, in effect, made the store clerk into a box maker. He had to mold the flat cardboard around a wooden form — the same procedure used in the box shops. The only real difference was that the box wasn't made until the customer actually purchased the tacks. This slowed sales; the basic point of packaging is to accelerate purchasing.

The first really useful folding box was a result of one of those happy accidents that play so prominent a role in the history and folklore of invention. In 1879, Robert Gair, a Brooklyn-based producer of printed flour, seed, and grocery bags, had been considering the problem of how to produce an efficient folding box in quantity. The insight came when a common mishap in letterpress printing occurred. A metal rule worked its way upward and began to make clean slashes in the bags Gair was making. This suggested to Gair that he could use sharp dies to cut the cardboard and, most important, blunt dies to crease the cardboard precisely. This also suggested that the box-making process and the printing process could be a single operation. Gair bought a secondhand printing press, on which he was able to produce 7,500 cartons an hour.

This invention led during the next two decades to nearly a thousand more patents involving the making of folding boxes and the creation of machinery to fabricate, handle, and fill them. Even more significant was the nearly instantaneous understanding of the distribution and marketing innovations this box made possible. Within only seven years of Gair's invention, Quaker oats, packing its product in one of these new boxes, was pioneering the principle that selling a product nationally in small, clean, distinctive packages could greatly increase demand for it. Within a decade, all sorts of factory-produced goods were being sold in folding boxes. And only a few years later, the Uneeda biscuit, the product that is synonymous with the end of bulk retailing, was introduced in its pioneering package with a sealed paper wrapper inside a folding box.

The folding box also gave rise to the most used shipping

container, the corrugated cardboard carton. Corrugated board had first been used in the mid-nineteenth century for the sweatbands of men's hats. In 1874 a process had been perfected for sandwiching a fluted medium within two sheets of board, and Gair's invention opened a vast new market for this material. Recent developments, particularly the use of paperboard that can support high-quality printing on the outside of the carton, has made these utilitarian containers increasingly packagelike.

Thus, despite the proliferation of packaging types and materials, paperboard still dominates, accounting for about 45 percent of the value of all packaging used and a larger portion of the weight and volume. In the United States, about six hundred pounds of paperboard are consumed per person each year.

In contrast with most other kinds of packaging, in which entrepreneurs and tinkerers were responsible for most major innovations, many of the chief developments of the canning industry were the result of direct government intervention. The long-term preservation of fresh food was conceived as a tool for empire. Missions of exploration and far-flung military installations were the first users, and war helped spur its progress. And in the United States, some well-timed government protectionism helped establish the industry and allow it to become dominant.

The technique of placing a thin layer of tin on the outside of a piece of iron to prevent corrosion was known in Roman times, and a center of tinplate crafting was established in southern Germany in the thirteenth century. The process required a long period of pickling the iron in a solution containing tin before the metal could be crafted into pots, serving utensils, canisters, and a wide variety of other useful objects. As the industry spread throughout Europe and tin supplies dwindled, producers began importing the metal from Cornwall. In 1706 England imposed a high protective tariff in order to establish a domestic industry. With ample Welsh coal and iron supplies, an industry that aggressively modernized production, and, later, tin supplies from its colonies, Great Britain dominated the industry for nearly two centuries.

Tinplate containers were appealing to some manufacturers because of their uniformity. Wooden kegs and barrels were all subtly different. Because of the varying characteristics of the trees from which they were made, it was not always possible to standardize their capacities. Large-scale production magnifies small discrepancies and has an impact on profitability. Moreover, inconsistent sizing jeopardizes a company's reputation for reliability. As early as 1813, the Du Pont Company declared, "We have been and are still using all our endeavors to have the whole of our barrels made perfectly uniform." While tinplate containers were also the result of craft production and thus inconsistent in some ways, they were produced from standard-sized sheets. Tinsmiths were aware of the most efficient ways in which a sheet could be used to make the most efficient containers. Thus, it was relatively easy to standardize the sizes of small, medium, and large tins, so that their capacity would not vary significantly. These sizes did not, however, correspond to pints, quarts, or round-numbered metric measures, which was a source of complaints and of manufacturers' concern for decades.

Tinware had a major role in food sales before the age of packaging. The containers in which grocers kept tea, spices, coffee, and other commodities were often made of tinplate. During the nineteenth century, there was a proliferation of tinware for the home, including highly decorated food storage containers and sealable shakers for certain spices and condiments. Some tinplate manufacturers began to sell special containers to grocers and to manufacturers who would pack their products in useful, decorated tins. Tin containers proved to be well suited for matches, at least until cheaper boxes and even cheaper matchbooks came along.

One of the most celebrated producers of decorative and novelty tins, the English baking firm Huntley & Palmer, represented an unusual synergy. The business got under way during the first half of the nineteenth century when Thomas Huntley set up a shop at a stagecoach stop at Reading. Apparently, the biscuits were good, and travelers began to take them away with them. Huntley asked his brother, a tinsmith, to make tins to pack biscuits for this itinerant market, and soon the biscuits were sold all over England and throughout the world. It is difficult to say whether they were in the cookie business or in the tin box business.

After the Somers Brothers of Brooklyn found the first practical process for printing directly on tin, decorative, collectible tin boxes proliferated throughout the world. Only a few years later, Lyons tooth powder, a product that seemed to demand a tin container with a reclosable top, made its appearance.

The technique of canning foods to preserve them was invented in response to a government-sponsored competition. In 1795 the Directory, at that time the governing body of France, announced a prize of twelve thousand francs for anyone who could come up with a new, effective means of food preservation. Nicolas Appert, a sometime confectioner, vintner, and pickle maker, theorized that if food could be cooked in bottles and kept from the air, it could keep. It's worth noting that this insight had no theory to back it up. Louis Pasteur's discoveries and the germ theory of disease were a half century away, and their practical application to canning was even more distant. Appert was working on the assumption that exposure to air spoiled food, just as it hastened the deterioration of the flavor of wine. Appert worked on this concept for the next fourteen years, aided somewhat by the French navy, which in 1806 tried out some of his meats, vegetables, fruit, and milk. In 1809 he decided he had proved his case, and the government's consultative bureau of arts and manufactures, after waiting a suitable period for Appert's food to spoil, agreed. Napoleon himself presented Appert with the prize, and he was said to be very impressed by the new mobility and healthier conditions such foods could provide for soldiers.

It did not, however, prove to be an effective military secret. In 1810, the year of Appert's patent, a patent was granted in England for a process virtually identical to Appert's, except that it used tinplate containers rather than glass jars. The first cannery using tinplate cans opened in Bermondsey, England, in 1812, and by the next year it was sending cans of its products to the army and navy for trials. In 1814 some of the canned foods were sent to distant British military bases, including St. Helena, where a few years later, Napoleon, the patron of canning, might have had a chance to sample them. By 1818 the Royal Navy was using about twenty-four thousand large cans each year on its ships. Other cans of what were sometimes called embalmed provisions went off on voyages to Baffin Bay and the Northwest Passage.

These cans consisted primarily of meat with gravy, meat with vegetables in gravy, and vegetable soup. They replaced at least some of the salt-preserved foods that had been the staple on long voyages and that often lost most of their vitamins and other nutritional value in the process. Although it was known that eating fruits and vegetables could prevent scurvy, there was also a persistent belief that salt meat is what caused it. The canned provisions were thus doubly attractive. Sailors who had gone months without seeing a vegetable could now encounter them more often, albeit cooked for hours as a part of the preservation process.

One of the early cans was a four-pound tin of roasted veal, manufactured by Donkin and Gamble, the pioneering English canning firm, for Captain W. E. Parry's third voyage to discover the Northwest Passage in 1824. It was a stout thick-walled container, bearing a label that read "Roasted Veal" in Gothic lettering and directions that among other things advised the use of a hammer and chisel for opening it. The can's notoriety stems from the fact that nobody aboard the HMS *Hecla* actually did open it. When, two years later, Parry set out again, it was once again loaded onto the ship. Aside from complaints that the vegetable soup was sometimes too sour, Parry and his ship's surgeon were strong supporters of canned foods and felt confident that the veal would survive a second Arctic expedition. Once again, the can returned unopened. It did not make a third voyage. Instead, it assumed the status of an artifact of Parry's quest and reposed in a museum until 1938, when it was opened for chemical analysis. "The meat itself was in what one could fairly call 'perfect' condition, the appearance of the larger fragments being quite like recently cooked veal," the researchers reported. They did note a surprisingly pink coloration and the presence of white fatty slime, but their chemical analysis showed that the can had held its nutritional value better than was predicted, and that while there was a good deal of tin to be found in the thin milky fluid in which it was packed, there was very little lead. It was fed to young adult rats for ten days, who were reported to have eaten avidly without ill effects, and one meal was fed to a cat, who was similarly unaffected.

More recently, however, discoveries related to another Northwest Passage voyage, the Franklin expedition of 1846–48, gave a sharply contrasting account. Both ships became ice-bound and were abandoned by their crews, none of whom survived. Fragments of the bodies of two sailors were recovered, frozen in the Arctic ice, and their tissues were found to have what researchers termed "catastrophic" levels of lead. This would have weakened them severely and certainly did not help their survival in a life-threatening situation.

The inconsistency of these two pieces of evidence about the survival of food in these early cans probably derives from the inconsistency of the cans. All of them were made by hand from heavy-gauge tinplate, in much the same way as a pot or canister. Beginning with fourteen-by-twenty-inch sheets of tinplate, they used tin shears to cut out the body and ends of the can. The side and end seams were soldered by hand, and a circular hole was left at the tip for putting in the food. After the containers were filled, heated, and exhausted, the canner soldered a tinplate disk at the top to close the hole. Lead-based solder was usually used to seal the seams at the outside, but differences in individual workmen or in the procedures of a specific shop probably made a difference in how much lead got inside the can. An expert can maker could do five or six an hour. In England, the canners seem to have employed tinsmiths; in Germany, making cans was a sideline for plumbers.

The consumer use of canning began slowly. In the 1820s fish canneries began to operate in Scotland. Thomas Kensett in New York and William Underwood in Boston, both recent immigrants from England, began canning operations during the first decade after Appert's technology was developed, and Kensett was granted the first American patent for preserving food in tin containers. He packed the first hermetically sealed oysters, meats, fruits, and vegetables in the United States. Underwood packed pickles and fruits in glass jars until, in 1837, a year of financial panic, he switched to tinplate to cut costs. Powdered milk was invented and canned in Russia during the 1840s, and salmon canning started in Ireland in 1847. But these remained relatively insignificant until some major technological advances at midcentury.

The invention in 1856–57 of the Bessemer steelmaking process had as a consequence the production of a new kind of tinplate, now mild steel rather than cast iron. This was more easily produced, more malleable, and better adapted to being in thinner sheets. In 1859 the firm of Wilson, Green and Wilson, working under contract to Du Pont Company to redesign gunpowder kegs, patented a tinplate container in which the ends of the sheet interlocked and were covered with a narrow strip to make the tie. No solder was used. This is the principle of the double-seam can, still in use today. Meanwhile, tinsmiths everywhere — especially in Baltimore, whose proximity to both seafood and agricultural products made it a canning center — were building die cutters, presses, and other semiautomatic devices that enabled them to increase can production to about sixty an hour in the 1870s. And in 1861 a Baltimore canner discovered that if calcium chloride was added to the water in which cans of food were cooked, higher temperatures could be achieved. This cut preparation times from six hours or more to about half an hour and immediately increased the productivity of all existing canning operations.

Such steady, incremental development of the technology of canning, especially in the United States, demonstrated a real demand for preserved, portable food. But technology alone could not have propelled a new way of preparing and keeping food into millions of households. It required a powerful personality, Gail Borden, who was determined to change people's lives — along with a series of events that led, during the 1860s, to a fivefold increase in American use of canned foods.

In 1846 shortly after the Franklin expedition began sailing about the Arctic, slowly poisoning itself with its canned goods, the eighty-seven members of the Donner party were trapped in deep snow as they were attempting to cross the Sierra Nevada and reach California. Unlike the Franklin expedition, in which all perished, forty-seven members of the Donner party survived, but only by eating the flesh of those who had already died. It would be 135 years before anyone would know of the role of tinned provisions in the Franklin expedition disaster. Meanwhile, the shocking tale of the Donner party focused the attention of Americans, and of one American in particular, on the need for better portable foods.

Borden seems to have always expected to do something that would change the world, though he wasn't sure just what. He was an inventor-promoter, whose one enormous success turned him from crank to genius overnight and made his obsessive rantings seem a great deal less ridiculous. Before he got into portable food, he was responsible for such devices as the locomotive bathhouse, a device that could be wheeled into the water to permit ladies to swim modestly. He also developed an amphibious, sail-driven wagon, intended to speed the pace of transportation across the continent. After this "terraqueous machine" sank during a demonstration, the fate of the Donner party prompted Borden to think about nutrition for travelers.

Borden was a transportation-oriented thinker, and his first instinct was to somehow shrink the food, so that there would be more nutrition in less space. This was not really a new concept. An English manufacturer had sold a highly condensed soup cake to the Royal Navy during the eighteenth century, and a gluelike chunk of it survived from Captain Cook's expeditions and was analyzed in 1938 by the same group that tested Parry's veal. It was considered to be useful in an emergency, but not really very good. Borden's version was a biscuit made from a concentrated extract of meat, combined with flour or vegetable meal and dried. Like the earlier version, it was meant to be mixed with water to make a broth, and like the earlier one, nobody wanted to eat it, given a choice.

Borden lost sixty thousand dollars on his meat biscuit, and most of six years of his life. But this failure did not cause him to lose his faith in condensing things. He believed that his age did not allow things to stretch out in time and space as they once had. A declaration of love was distilled to a kiss, a sermon to an aphorism. A poem was a luxury nobody could afford. He boasted that while people had once taken hours to eat dinner, Napoleon did so in twenty minutes, and he had cut the time to fifteen minutes. "I mean to put a potato into a pillbox, a pumpkin into a tablespoon," he declared.

What he did do was put condensed milk into a can. He wasn't the first to have canned milk. Appert had done that, and so had others since. His vacuum-canning process for milk was new, or so he convinced the reluctant U.S. Patent Office. Using a vacuum pan he first saw at the Shaker colony at New Lebanon, New York, Borden had the inspiration of greasing the pan to prevent the milk he was boiling from sticking and foaming. This simple home remedy worked, and Borden was able to evaporate much of the water from the milk and then can a product that many people found palatable. Earlier canned milk products did not really taste very much like milk, while Borden's seemed to have been close enough to be acceptable.

We can't know now whether the taste of Borden's milk was much better than that of his predecessors. His promotion was, however, outstanding. With an investor, he was able to set up a cannery in Connecticut and target the New York market. This was a far cry from outfitting the Donner party, but Borden's advertising was up to the challenge. Rather than describe the milk as a portable convenience, his advertising positioned it as a product that was cleaner, purer, and fresher than anything that was otherwise available to city residents. Thus, Borden was surely among the first food producers to make the fundamental claim that packaged and branded goods are safer and cleaner than others. Like many merchandisers who came after him, Borden presented a vision that fused sun-drenched clover-laden pastures with modern techniques and high standards. (Much later, a competitor of Borden's company even claimed "contented cows.") This pastoral-hygienic image stood in stark contrast to what people saw all around them: dirty stores with flies buzzing around the milk bucket, or wagons on which uncovered containers of milk caught the dust and filth of the manure-strewn streets. The reclosable milk bottle wasn't invented until 1886.

Goods in packages are not inviolate, but they do tend to be

cleaner than those sold in bulk. Packaging a product limits the amount of time it is subject to abuse. Moreover, the producer who has invested in a large processing plant faces the risk that a few small failures will damage his reputation. Because this is a possibility the processor wants to avoid, hygiene and quality control are major concerns of big operations.

Borden's plant was a hundred miles away from New York, and the farmers he dealt with would probably not have had any access to that market. By increasing their market reach, and relieving them of the necessity to find outlets for their production, Borden had the leverage to demand that the farms operate in a clean way and deliver the milk to the plant cold.

Borden's entry into the New York market in 1858 coincided with a milk-supply scandal. Many cows on farms close to the city were being fed on slops collected from liquor distilleries, and their milk was found to have very low nutritional value. Such low-quality milk was often adulterated in order to give it the appearance of milk. Reformers, including *Frank Leslie's Illustrated Newspaper,* where Borden advertised, contended that bad milk was a major cause of malnutrition and the deaths of children. Borden's potential customers were literally frightened into trying something new. And what he had to offer was, in fact, a wholesome, nutritious product.

The event that transformed Borden's business, and that of all the other canners who were operating at the time, was the Civil War. Borden couldn't keep pace with the Union army's demand for cans of his milk, so he licensed condensed-milk plants elsewhere throughout the country. Even while the war was going on, the demand from these plants on farmers changed the nature of the dairy industry, as more and more farmers opted for the guaranteed income of a contract with a milk processor.

The war increased demand for canned vegetables and fruits from other suppliers as well. At least in the North, a generation of soldiers became accustomed to eating canned foods. This was hardly a minor matter. Just before the war, American output of canned foods amounted to about five million cans each year. Only a few years after the war, this figure had grown to thirty million.

Moreover, as World War II was to do more dramatically, the

71

war increased demand for metal and increased creativity and efficiency in can production. That led to thinner but still sturdy cans that used somewhat less tin, a resource the United States lacks. And in 1865 came an invention that truly removed one of the major obstacles to people's use of tinplate containers — the can opener.

In 1862 Louis Pasteur published his research that showed why the canning process worked, but this really had little immediate impact on the industry. Pasteur's ideas obviously had impact on the processing of fresh milk, and they underlie today's aseptic packaging, in which contents and container are sterilized separately and filled in a germ-free environment. And in a more general sense, his implication that touching or breathing on food spreads disease undoubtedly had an impact in winning widespread acceptance for packaging.

But the canning industry had learned Pasteur's lessons without knowing it, and, since the invention of the retort, or pressure cooker, in 1874, there has been little conceptual change in canning. As in nearly every other industry, the late nineteenth century brought the canning industry tremendous advances in productivity through technologies that mechanized craftsmanship and optimized product flow. By 1890 semiautomatic processes increased the production rate to 2,500 cans an hour. A decade later, production was 6,000 an hour, and many of the companies that remain household names in the canning industry were already well established.

One important American development of the 1890s was more political than technological. Americans had become the world's best customers for canned goods, and because of the country's vast growing regions and large fishing fleets, it had also become an important exporter. Chicago, with its stockyards, had become an important center for canned meat. But the United States had no tinplate-producing capacity and imported all the material for its cans from Britain, which controlled tin supplies in Malaya. In 1890 Congress, with the specific goal of creating an American tinplate industry, imposed a heavy tariff on imports of the material. This enabled Americans to purchase tin from the Congo, Bolivia, and elsewhere and begin their own industry, which by 1912 was the largest in the world. The building of U.S. mills coincided with and very likely helped spur explosive gains in the production of cans and of consumption of the products. At the same time, the production of cans became a separate industry from the packing of food. In 1900 a financial syndicate purchased 123 can companies, representing 90 percent of American production, and formed the American Can Company, which long dominated the American industry. In 1941 *Fortune* cited American Can as one of the five corporations that "more than any others shaped the daily life of man in the United States."

The two decades that straddle the turn of the century were filled with innovations in the packaging industry as a whole. In 1892 the metal tube, which had been invented during the 1840s for holding artists' colors, took on a whole new life when Dr. Worthington Sheffield, a New London, Connecticut, dentist, had the idea of filling it with toothpaste. He quit dentistry and went into business making tubes for others. Only a few years later, Colgate, the venerable New York druggist, adopted the practice, and by the first decade of the twentieth century, a wide variety of ointments, creams, and other personal-care products could be squeezed, hygienically and in small quantities, from tubes.

Changes in one of the oldest container industries, glass, were particularly dramatic. While there had been advances in the efficiency of furnaces as the nineteenth century began, the procedures for making a bottle had not much changed since Roman times. Only in 1821 were molds introduced that allowed the body and the neck of a bottle to be blown together. This led immediately to a lot of fancy, decorative bottles. One particularly well-known Philadelphia-made whiskey bottle featured the face of George Washington in relief on one side of the bottle and the bottle maker, an Irish-born entrepreneur later jailed for fraud, on the other.

But until almost the end of the century, all bottles involved lung power. Experienced men blew the bottles, boys held the molds and tore the bottles off the pipe, and other craftsmen finished the necks and tops. After 1861, gasified coal and later natural gas added to the efficiency of the glassmaking process, and the screw top, invented in 1872, made bottles and jars more useful than they had been before. Nevertheless, a state-of-the-art bottle shop of 1880 consisted of three men and four boys. Handling molten glass is difficult and dangerous and thus was long resistant to mechanical simplification. Semiautomatic devices, which formed the container but left the gathering and feeding of the molten glass into the machine to craftsmen, were invented during the 1880s and came into widespread use during the 1890s. Michael J. Owens's fully automatic machine, based on a suction principle inspired by the bicycle pump, was patented in 1903.

The complexity of the Owens machine is almost a sufficient explanation for why it was so long in coming. It consisted of several complete working units mounted on a continuously rotating framework. With each revolution, each unit produced a complete bottle. Each unit contained separate molds for the body and neck of the bottle. Using air pressure, glass was pushed into the molds, then evacuated through holes within the molds, and the glass clung to the outside. Knives trimmed the tops and bottoms. After 1917 new machines that permitted a continuous flow of molten glass supplanted the Owens device.

Perhaps the most surprising thing, in retrospect, about the slow development of bottle-making technology is that for some industries it didn't much matter. In England, especially, the nineteenth century brought an enormous number of bottled sauces. In America, H. J. Heinz, after failing in a venture to pack horseradish in clear bottles to prove it wasn't adulterated, began in 1876 to pack his ketchup. Liquor, wine, and beer were important bottled products nearly everywhere, and Bass ale was, in 1875, the first trademark granted in Britain. The emergence of St. Louis as a railroad center and the invention of the crown bottle cap in 1892, which enabled beer to keep better, suggested to Anheuser-Busch that it would be possible to create a nationally distributed beer. In 1896 the company placed an order for half a million caps, one hundred thousand immediately and fifty thousand a week thereafter. Slow bottle making does not seem to have been an obstacle.

*   *   *

The creation of a delivery system, able to transport the heavy bottles of Budweiser throughout the continent at a price that could compete with local brews, was far more important for Anheuser-Busch than the mechanized production of the container itself. While knowledge of the evolution of different sorts of containers contributes to an understanding of packaging, it is important to remember that packaging is more than containers. It is also a means of conveying information and changing and sometimes circumventing human relationships. Packaging changes the way people understand the world.

Its first effect is to weaken people's dependence on the place where they are and make them more dependent on a large-scale delivery system. As we have seen, canning sprang directly from French and English imperial ambitions, and the technology was of great interest in the United States and Russia, where there were internal frontiers to conquer. Archaeologists, after studying the garbage piles from saloons in old western ghost towns, reported that the cowboys and the miners really like smoked oysters, though they were thousands of miles from the nearest living bivalve. On the frontier, or in the colonies, most things came from somewhere else, the settled places, the metropolitan centers. Soon, of course, some things could be processed in remote areas and sent back. For example, the Pacific salmon-canning industry was well established in California a mere fifteen years after the gold rush, and by the 1890s can manufacturers were making special collapsible cans to be shipped to Alaska to be filled with salmon.

The experience of Gail Borden, who set out to prevent more Donner-party catastrophes and ended up selling milk in New York, suggests that the rapidly industrializing cities were themselves frontiers of a kind. Philadelphia, whose population was 400,000 in 1860, quadrupled to 1.6 million by 1910, and several other American cities showed comparable growth during the same half century. During the two decades from 1890 to 1910, American population increased by more than a third, from about 64 million to 88 million, with growth heavily concentrated in the cities.

Many of these new residents were immigrants, many of whom had left a village life of complex social relationships and regular, intimate contact with their food supply. They had forsaken their aunts, their uncles, and their cows and subsistence economy for the work-centered, wage-earning, often anonymous life of the industrial cities. Trying oatmeal or canned milk was not any more drastic than what they had already done.

Thus, by the end of the nineteenth century, there were new immensely productive ways of making things and transporting them, technologies for making containers rapidly and cheaply were advancing quickly, and there was a large population receptive to change. There was a lot of stuff, and people who might buy it. But this was not enough. In order to protect their enormous new investments, manufacturers had to find some way to assure that people would not only buy things, but that they would be able to find their products and purchase them repeatedly. And to accomplish this, they turned to the tactics of a business that had, for centuries, proved that it could sell unknown ingredients to a loyal public — patent medicines.

The adaptation of the insights of patent-medicine packaging and marketing to food, drink, and nearly everything else an individual could buy was a relatively late stage of industrialization, a key step in making the purchase and use of industrial products keep pace with their production. Meeting pressing needs that buyers could understand was generally not enough to keep the machinery turning at its optimum efficiency. Instead, producers began to focus on desires and dreams. The way to spur the consumption of material goods was to dematerialize them. A food was not just something for breakfast, but rather something to inspire confidence, an indication that you were doing a good job as a mother. And, as we shall see, these same intangibles were very important to manufacturers, who needed to be able to predict their production flow, and not be vulnerable to purely price-based competition. If what you're selling is not just cereal, but an expression of maternal love, it can survive a few pennies' rise in grain costs.

It is appropriate that the first product to be packaged and marketed in a way that we can recognize as wholly modern was so uncharismatic a substance as oatmeal. Those who knew about it at all thought it was mostly for horses and a few stray Scots. Yet what was animal feed in 1870 was effectively marketed two decades later as "a delicacy for the epicure, a nutritious dainty for the invalid, a delight to the children." What had happened in those twenty years was alchemy through packaging. A base substance had been put in a small box, invested with personality, outfitted with recipes that increased its usefulness, and turned, perhaps not to gold, but to something that was desirable and profitable. And within years after the Quaker appeared on the box, the package design was mobilized for a wide range of marketing gimmicks — cents-off promotions, miniature sample-sized boxes, box-top premiums, coast-to-coast publicity campaigns. Many of these have their roots in the selling of patent medicines, but their integration with a national distribution system and the development of advertising transformed them from intimate little confidence games into cornerstones of popular culture.

Both these advances — the ability to distribute the product and the ability to add information and personality to the product — were essential to the success of Quaker and all the packaged goods that followed. But they did not happen simultaneously. Mass production and distribution came first, and the problem of inducing and managing consumption was not understood until sometime later. The rise of oatmeal required two geniuses, the miller, who turned oats into a popular commodity, and the promoter, who turned the commodity into a far more popular product.

Ferdinand Schumacher, who built his first mill in 1856, is generally credited with introducing oatmeal to the United States as a food for humans. His first customers were the hordes that were immigrating from Ireland from the midcentury onward and some German immigrants like himself. He first shipped his product in small glass jars, but as demand picked up, he moved to less expensive, more durable barrels, from which the grocer would sell in bulk. He forestalled the danger of oversupply by demanding that wholesalers place their orders for the entire year immediately before the harvest, so that he would know exactly how much grain to purchase. The oatmeal barrel never became quite so ubiquitous as the flour barrel and the cracker barrel in American groceries, but it became sufficiently widespread by 1880 for the *New York Times* to call it "a craze." More concretely, the demand became great enough to justify Schumacher's building of one of the largest mills yet constructed, his five-story Jumbo Mill in Akron. When this mill burned in 1886, a significant amount of the country's oatmeal-producing capacity was destroyed.

At this point, Henry Parsons Crowell entered the picture. He owned a small mill in Ohio. But in contrast to Schumacher, a man truly devoted to oats, Crowell believed that God had turned his attention to oatmeal as a means to a larger end. When he was a young man, he said, he believed that he was about to die from a heart ailment. At that time, he made a pact with God that if he lived, he would give his money to support religious evangelism. He upheld that bargain by contributing 65 percent of his considerable income from milling and other interests to Protestant causes. (He does not appear, however, to have given money to the Religious Society of Friends, the Quakers.) Like another religious businessman, Andrew Carnegie, Crowell had ambitions as a monopolist. From the time he got into oatmeal, he began thinking about how to assemble a combination in the grain-milling industry similar to that Carnegie had assembled in steel. Before the Jumbo Mill fire, Crowell and some allies had sought to create a pool to fix prices in the industry, but Schumacher had doomed it by refusing to join.

Crowell's immediate responses to Schumacher's fire were to greatly increase his oatmeal production and to transform the way it was sold. He made use of the cardboard folding box Gair had patented only seven years before and decorated it with the figure of a rather austere Quaker. There was also a recipe for how to cook the oatmeal. Crowell had ample precedent for using the Quaker from patent-medicine selling. He also might have been inspired by the plan, announced that year, to place a colossal bronze sculpture of William Penn atop Philadelphia's city hall. Although there is scant facial resemblance between Crowell's package and Alexander Milne Calder's statue, it is striking that both are standing figures holding scrolls in their left hands. Calder's Penn is holding the charter of Pennsylvania; Crowell's Quaker's scroll bears the word PURE. Crowell went to a stouter, friendlier-looking Quaker a few years later.

Schumacher's fire also afforded the opportunity to reassemble the price-fixing pool. But because its members accounted for only about half the industry's capacity, the pool could not control production, and it collapsed. Two years later, seven of the largest millers, including both Schumacher and Crowell, joined to form the American Cereal Company. At first, each of the millers who had formed the company operated independently. Schumacher continued to sell his oatmeal in 180-pound barrels, while Crowell introduced a four-color package for Quaker oats. The recipes on the package became much more extensive, inviting housewives to use the product to make bread, fried pudding, pancakes, and baby food, as well as breakfast porridge. The copy on the package was decorous but highly promotional: "We would call your special attention to the purity, rapidity of preparation, and the fact that they did not sacrifice the sweetness and flavor for the sake of rapid cooking."

One of his partners developed an automatic machine, which could assemble and fill twenty boxes of Quaker oats per minute. The machine also enabled him to put free spoons in the boxes to spur more sales. He began to advertise. "One pound of Quaker Oats makes as much bone and muscle as three pounds of beef," his first newspaper ad said. "Is it worth trying?" He was also successful in convincing publishers to print advertisements that were disguised as news stories, all of which pointed to the remarkable properties of his oatmeal.

In 1891 he chartered a special fifteen-car freight train that operated like a heavy-metal medicine show, running from Cedar Rapids, Iowa, to Portland, Oregon, stopping frequently for the distribution of miniature packages of the product. The idea was to create demand for the oatmeal before Quaker salesmen called on grocers. The knowledge that there was actual customer demand for the product helped break down grocers' resistance to selling the oatmeal differently and being tied to a particular supplier. He sought to reinforce recognition of his emblem by offering premiums shoppers could get by cutting the Quaker off the package and mailing it in. Quakers were painted on fences, the sides of buildings, inside streetcars, and on billboards. In the space of only a few years, Quaker oats became the most promoted product ever.

But Crowell was still the partner of a man who believed that advertising and promotion were a waste of time. Schumacher was the largest holder of American Cereal, and he succeeded in getting Crowell and his allies forced out. A few years later, Crowell regrouped and, in a pioneering hostile takeover of a company, forced Schumacher out.

Crowell's return did not, however, bring any major new marketing innovations. In the space of only a few years in the late 1880s and early 1890s, he seems to have exploited the idea of a packaged good in every way possible. And by the turn of the century, many others had learned from him.

4 *The grocer couldn't help it*

*I*n 1901 Uneeda biscuits ran this small masterpiece of advertising copywriting in dozens of newspapers and magazines:

> The Grocer Couldn't Help It!
> When the grocery boy swept out the store, he couldn't find the lid of the cracker box, so he covered them over with a codfish crate. After he weighed a mackerel and drew a quart of kerosene, he filled an order for a pound of crackers, which was carefully picked from the cracker box under the codfish crate. The customer who bought the crackers said they tasted queer, but the grocer couldn't help it.

The only other thing in the advertisement was the orb-and-cross logo, a variation of which is still used by Nabisco. This symbol, which had been adapted from a fifteenth-century Venetian printer's mark, had previously been used as a symbol of Christ's redemption of the world. Now it was used as the mark of Uneeda's patented "In-Er Seal" package. It symbolized the cleansing of the cracker and the raising of the grocery buyer's expectations.

There has probably never been a product that sold its package more aggressively than Uneeda. Early advertising for the product, which was introduced in 1898, barely mentioned the taste of the cracker. It concentrated entirely on its preservation in a double-sealed package — a folding box on the outside and an inner lining that was also attached to the box. There was nothing new about the cracker, except that it was clean and crispy, free from insects, untouched by rodents. The promise was that it was what crackers from the barrel would be like if you took them from the oven before they were subject to abuse by grocers. Packaged crackers weren't new; British bakers had been putting biscuits in tins for decades. Nevertheless, the saturation advertising campaign was so strong and effective that nearly a century later, Uneeda is widely thought to be the first packaged product.

Uneeda's grocery vignette, which would be farcical if you did not have to imagine eating the results, performs the difficult task of dramatizing an absence. The reason for buying a box of Uneeda was what wouldn't be in it. The purchaser could be certain that no floor sweepings, salt cod, mackerel, or kerosene had found their way into the doubly sealed packages of crackers.

At the same time, the tone of the writing is gentle, and it purports to be understanding. The grocer's boy was the unwitting agent of all the sanitation problems. The advertisement seemed to assume that the reader probably liked and trusted the grocer, who was very likely well meaning. But, it hinted, good intentions are not enough to keep something fishy from happening to your crackers.

Packages change the way things are sold, which is precisely why retailers resisted them. The Uneeda advertising campaign was a delicate attack. It did not seek to sunder shoppers' relationships with people they trusted, their grocers, but it sought to introduce them to something they would be able to trust even more, packaged, branded, advertised products.

"With the forefinger of your right hand," an earlier Uneeda ad advised, "point to the shelf and say to the grocer 'I want these biscuit [sic] in the In-Er Seal Patent Package.' Then he'll know that you know that the cracker barrel and the paper bag are out of date." What the makers of Uneeda really wanted, of course, was for the grocer to know that the box had supplanted the barrel, and his store would never be the same. The saturation advertising of Uneeda biscuit, and the particular emphasis it placed on the qualities of the package, represented an intensification of such Crowell promotional efforts as the special train and miniature free samples.

The struggle to turn commodities into products involved protracted guerrilla warfare. Salesmen could go door to door, selling or distributing samples of Comet rice or Argo cornstarch in bright, memorable packages. They told buyers to ask for them when they went to the grocery store. Then they went to the grocers to try to make the sale, which was usually harder. They had to convince the grocers that they lost more money spilling rice from the barrel than they made by short-weighting it for their customers. They told them that they would not need so much help in the store to wait on customers and that the store would be easier to clean. But thwarting the packaged-goods salesmen were the jobbers, whose relationships with the grocers were long and established, and the grocers themselves, who realized that they were losing a measure of independence and did not understand what they would get in return.

The goal of the manufacturer of packaged, branded food was to use the ultimate consumers of the product to put pressure on grocers to stock the product, and for the grocers to put pressure on wholesalers. The package represented a direct challenge to the wholesalers' power over food processors. Wholesalers could always forsake a manufacturer of bulk items if someone else offered the same product on more attractive terms. But a packaged item allowed manufacturers to defend themselves against the natural adversaries on whom they depended to sell their goods.

It seems as if the power of modern marketing should have swept the messy barrels and bulk sales out of the store, but it didn't happen. The struggle between the barrel and the box lasted half a century. Only during the 1930s did packaged brands manage to conquer such bulk-goods strongholds as sugar.

Packaging, branding, and advertising are obviously not synonymous. Countless canneries produced very attractive cans that emphasized the quality of the beans or peas inside but did not establish the identity of the canner. Schumacher, by contrast, managed to establish a modest level of brand identity for his oatmeal without either packaging or advertising. And Milton Hershey's chocolate bar became a powerful brand, and its package an icon, without any use of advertising. (The first Hershey advertisement was done in 1970.)

Nevertheless, when packaging, branding, and advertising began to be used together, the impact was greater than the sum of their parts. Branding is the manufacturer's assertion of standardization and quality. Packaging is both a physical protection that enables the manufacturer to maintain control over the contents, and thus assure its standards, and an expression of identity. Advertising lets people know about the manufacturer's promise and makes them familiar with the package.

Crowell had integrated these three marketing tools in order to turn a relatively unfamiliar commodity into a popular product. But like the patent-medicine promoters he emulated, he sought to sell his product for a wide variety of uses, as an ingredient in pancakes and bread, and as baby food as well as breakfast cereal, itself a new category of use. Such variety required that the consumer have a good deal of information about the different ways the product could be used. That's why he covered the package with recipes. But he did not understand that one of the great advantages of packaging is that it can simplify the way in which people use a product. He established Quaker oats as a brand and as a product, but he did not really understand the difference between the two.

One who did was Gerhard Mennen, a Newark druggist. He was a German immigrant with an engineering bent who went into pharmacy after he was weakened by malaria while surveying the swamps of Hoboken. His talent was for product invention and marketing. His first foray into both came in 1878 when he developed Mennen's Sure Corn Killer, which he sold through a show that involved black banjo players, whose music was punctuated by his own lectures on corns. He did not, however, attempt to sell on the spot, but instead sent potential buyers to local pharmacies. By doing this, he probably forfeited immediate sales, but he was able to build a far larger marketing network. And although Mennen came directly out of the medicine show tradition, he objected to multipur-pose cures. He sought treatments for conditions that were very specific, yet widespread enough to be profitable.

He turned his attention to the chafing of babies' skin, a condition for which there were many remedies on the market, none of them very effective. In looking for something slippery that stayed dry, he learned of a kind of talc mined in the Italian Alps. He ground this, medicated it with boric acid, and added oil of roses for scent. He first marketed it in 1889 in a round paperboard box with a pinwheel top, but he soon changed to a leakproof container of tinplate with the same sort of shaker used for spice containers.

Working with the Somers Brothers of Brooklyn, the inno-vative can manufacturers, he ultimately came up with a printed enameled can with a patented closure that permitted the powder to be shaken, and then the can could be snapped securely shut. Men-nen marketed this as "the box that lox." Mennen and Somers also perfected techniques such as bottom-filled cans with nondetachable tops that would not permit refilling with inferior or adulterated contents, and a double-sealed can that kept the contents secure and free from moisture. Mennen's advertising identified the integrity of the cans with that of the man whose face was on them.

Like many manufacturers of his time, Mennen decided to use his own picture as his trademark. It was assumed that nobody would be brazen enough to pirate the owner's own face. But instead of putting this mark on the front of the package, Mennen placed it on the top, where a potential customer could check it for authen-ticity. A mother from Akron wrote Mennen: "My Florence is so grateful that she kisses the picture of the gentleman on the lid all day long." On the front, he placed another picture, that of a happy baby, based on a photograph of the child of one of Mennen's sup-pliers' salesmen. Many previous packages had featured pictures of the manufacturer, of his store or factory, of the contents of the package, or of medals the product had won. Mennen's smiling baby was one of the first to show the happy result of using the product.

But perhaps Mennen's most important innovation was what would now be termed line extensions. Mennen knew that his talc would be useful to others besides babies. Nevertheless, he chose to identify it very specifically and establish confidence in his name. After a mother had entrusted her baby to Mennen, she was well disposed to try whatever else he might offer. He was soon able to make slight modifications of the product and market them as Violet Borated powder and Sen Yang toilet powder. (A men's powder was introduced in 1902, immediately after Mennen's death.) Thus Mennen had created a true brand — an established, trusted family of different products that answered individual needs and desires.

As we have seen, food canners processed whatever was available at any given time, which meant that their output was extremely varied. But for a very long time, they did not use packaging and advertising to establish brands. The multiplicity of interesting trade names on the can labels, with their illustrations of fruits and vegetables, country scenes, and canneries, had unmistakable nostalgic appeal. Moreover, they were, in general, more attractive than the branded lines that followed them. But most canners concentrated on selling each batch, rather than establishing a clear, year-round identity for their products.

One of the first attempts to do so was the introduction in 1878 by J. A. Wilson, the Chicago meatpacking firm, of a tapered, flat-sided can. The user would open the wide end, tap on the narrow end, and the meat would come out in a single piece for slicing. A few years later, this package was imitated by Libby, McNeill & Libby. Wilson was unable to protect the distinctive shape of its can, a result that encouraged others, including Mennen and Uneeda, who made packaging innovations that were functional as well as expressive, to patent their containers.

Probably the most successful use of a package to establish brand identity for canned goods was Campbell's soup. The company, which had been a regional canner buying from south Jersey farms and selling in the Philadelphia area, decided in 1899 to go national with its line of condensed soups. The advent of high-speed can manufacturing and food-processing and filling machinery had greatly increased canners' investment. Their activity could no longer be seasonal; profits depended on keeping the machinery running all the time. That virtually required that they have a wide range of products to can, many of them unfamiliar to consumers. The urgency of continuous-process production spurred important innovations. Tuna, for example, was first canned in 1909 by a California canner who had run out of sardines and needed something to keep production going.

For Campbell, the urgency was even greater. Most food processors were shifting to purchasing their cans from specialized companies like American Can, but Campbell stayed in the manufacturing business and eventually became one of the world's largest producers. As millers had discovered earlier, success in speeding production creates a new challenge: to convince people to buy all the new stuff.

Campbell had previously had a cluttered label and a number of brand names. Campbell sought to protect itself from its many competitors by creating both a unique product — canned condensed soup — and a very strong identity. Just as the soups were concentrated to save transportation cost and space on the shelf and in the kitchen cabinet, so was the expression of the can. All pictures were removed and the Campbell name was highlighted. The red and white was suggested by somebody who had recently attended a football game between Cornell University and the University of Pennsylvania and liked the look of Cornell's uniforms. The gold medallion represented a gold medal the product had won at the 1900 Paris Exposition Universelle. And that's about it. Today, when package design decisions involve many different corporate executives and outside marketing and psychological consultants, it's not likely that one man's ruminations at an Ivy League football game would be allowed to determine the design. For its time, however, the design represented a radical simplification, a move away from the visual complexity of similar products. Besides, red and white are an effective color combination, especially if your goal is to get attention. In conjunction with extensive advertising on trolley cars, billboards, and in magazines and newspapers, this new can established not only a brand, but a new necessity of everyday life.

The relative austerity of Campbell's packaging did not carry over into its advertisements. These frequently featured characters that played an important role in the company's success — the pair of pudgy, apple-cheeked youngsters known as the Campbell Kids. Lively, benign, and visibly well fed, the kids appeared as paper dolls, dolls, and other items that promoted the soup. As one observer expressed it, the Campbell Kids "not only sell Campbell Soups but help you to enjoy it."

The use of human characters in advertising and on packages marked an important shift in people's understanding of products. It was an indication of manufacturers' understanding that packaged, branded products had to establish a relationship with the consumer, a relationship that replaced, or at least supplemented, any relationship that the buyer might have had with the storekeeper.

Many early, preindustrial packages had sought to establish a geographic identity by printing addresses prominently and even making the labels look like the store signs. They evoked the act of going to a particular place to buy a product. Gradually, this gave way to the practice of wrapping the product in authority. In Britain, labels claimed royal or noble patronage and featured coats of arms. Elsewhere, canners, tobacco merchants, and match manufacturers invented heraldry to make their names more impressive. The proliferation of international exhibitions and trade fairs during the second half of the nineteenth century left many products festooned with more medals, ribbons, and badges of honor than the greatest war hero.

When places were depicted on nineteenth-century packages, they were rarely the store where you would buy the product and more frequently the factory in which it was produced. One late-nineteenth-century American canner, anticipating contemporary "green" marketing, showed his cannery humming away, with people in the foreground catching trout from a gurgling stream. An American seller of stock labels and bottles, which were sold to local distilleries to give them a branded look, offered a beautiful label showing an imaginary distillery, from which the whiskey was supposed to have come.

Packaging design follows fashion in design and typography,

but usually by quite a distance. Letter styles that were popular in the 1870s appeared on new packages in the 1890s and, in a few instances, survive today.

In the use of commercial characters, however, late-nineteenth-century packaging was in tune with the psychology of the times and was probably a leader. As the little boy in the raincoat carrying his box of Uneeda showed in countless advertisements, packaging is a way of defying the immediate environment. Earlier, packages had been viewed as parcels from somewhere else. Through the use of characters, they were newly understood as friends who could be found nearly everywhere.

These commercial characters could be fictional or generic — such as the Quaker or the mother on the oatmeal box or the ebullient, fat Aunt Jemima on the ready-mix-pancake box. It could be a real but oblique reference — the Log Cabin cabin-shaped pancake-syrup tin was a conscious effort to capitalize on the trust many Americans felt in Abraham Lincoln without offending Southerners. Most often the figures of trust were the manufacturers themselves, such as the Smith Brothers, with the words "Trade" and "Mark" beneath their pictures, or Gerhard Mennen, the face to be trusted in talcum powder.

The commercial characters emerged at the same time as such other popular cultural phenomena as the comic strip, and for some of the same reasons. Color images that were cheap enough to be disposable happened on packages long before the first comic strip, "The Yellow Kid," began appearing in the *New York World* in 1895. But the kid, who gave his name to yellow journalism, was soon enlisted for cookie boxes and other products. Both branded products and cheap newspapers responded to a new mass-consumption society.

The faces on packages were part of what the cultural historian Warren Susman has termed the transition from judging people in terms of the unchanging, duty-based attribute of "character" to the more aggressive, self-presentation-based notion of "personality." He says that, at the time, personality was almost invariably described in terms of the ability to stand out in a crowd, something that both people in an urbanized society and packages on the shelf need to do. Character is visible only over time to people living within a more or less stable community, but personality is how strangers get to know one another.

Increasingly, people understood that the new frontier of material progress consisted not of obtaining land but of winning the esteem of the strangers all around. "Young man, make your name worth something," the steel magnate Andrew Carnegie advised a graduating class of the Stevens Institute at the turn of the century. In an earlier time, this might have been an exhortation to character in an Emersonian vein. But Carnegie explained that he had something else in mind: "If you can sell a hat for one dollar, you can sell it for two dollars if you stamp it with your name and make the public feel that your name stands for something." In other words, your name is worth something only if people are willing to pay extra for it. In 1905 *Printer's Ink*, the publication of the rapidly changing and expanding advertising industry, estimated the value of one good name, Royal baking powder, at five million dollars, "a million dollars a letter."

One commercial character who survives, albeit modified, from the turn-of-the-century packaging revolution is Aunt Jemima. She has always dominated the product's packaging, and she seems, at first glance, to be simply a variation on Crowell's famous Quaker. Indeed, she has been a Quaker Oats Company brand since 1926. But she played a very different role in the history of consumer marketing.

With his brilliant marketing of Quaker oats, Crowell had transformed a relatively unfamiliar commodity into a popular product. The Quaker is a trustworthy figure to inspire confidence in something new. Aunt Jemima, by contrast, appears on a package of commonplace ingredients, mostly flour. Pancakes had been widely eaten for many centuries before she first appeared. What was new about the product was that the ingredients had been combined and were presented as foolproof. Aunt Jemima embodies convenience, a vision of ordinary life made easier. This is the promise of many, if not most, packaged products.

In 1889 Chris Rutt, a St. Joseph, Missouri, newspaperman who had unwisely bought a flour mill, did not see any way of profiting from just selling his flour. He was at the mercy of giants who could manipulate the market and force him out of business at will. His only hope, he decided, was to use his ingredient as a component of a product that could be marketed as something special. If he could establish the product, he could protect the market for his flour and make a higher profit from it. After much experimentation, he perfected a ready-mix pancake batter.

While he was working on the formula, Rutt attended a local vaudeville house where the performers included Baker & Farrell, white men who performed in blackface. One of their numbers was "Aunt Jemima," a cakewalk that featured Baker in an apron and red bandana. Much has been written about the metamorphosis of Aunt Jemima's image over the years, but it has been little noted that she started out as a white man in drag. Nevertheless, Rutt viewed the bandana-clad black cook as "southern hospitality personified" and used the name and image for his product. But he was undercapitalized and could not raise the money to do the marketing campaign the product needed, so he sold it to another company, which revised the formula by adding powdered milk to make it even easier to use.

The new owners decided that there should be a living Aunt Jemima, and after a search, they found Nancy Green, who played the role for more than thirty years. She made her debut at the Columbian Exposition in Chicago in 1893, where she unveiled her slogan, "I'se in town honey." The company published a souvenir booklet, *The Life of Aunt Jemima, the Most Famous Colored Woman in the World.* In 1895 the product began to be distributed in folding boxes rather than paper sacks, which gave greater opportunities for store display and a medium for paper dolls that increased the product's popularity with children. Green herself made up much of the legend of Aunt Jemima in answer to questions, and eventually the company had the eminent illustrator N. C. Wyeth do a series of advertisements showing dramatic scenes from Aunt Jemima's life.

African American characters, such as the Cream of Wheat cook and the big-lipped, bright-smiling Gold Dust Twins, were prominent in early packaging and advertising. These racist images, and others that reflected ethnic stereotypes, provided products with personalities that were apparently unthreatening. They were a kind of servant just about anyone could afford, with the frequent exception of people who belonged to the groups shown on the package.

Today's Aunt Jemima, even in her upscale guise, is something of an anachronism, since most of the commercial characters that arose at the turn of the century have disappeared, and relatively few have taken their place. But Rutt's key insight of increasing profit and insulating himself from the vagaries of the grain market by combining common ingredients in a useful and memorable way remains valid. It is the rationale for thousands of the products found in every supermarket.

Such new kinds of products challenged old kinds of grocery stores, and they helped to produce new kinds of consumers.

Perhaps we shouldn't accept the unavoidable, and unappetizing, comedy of errors depicted in the Uneeda advertisement as a wholly accurate account of the turn-of-the-century grocery store. For one thing, Uneeda's grocer seems to have dealt in more fish than appear in other accounts. Moreover, some packaged items, notably such high-value imports as chocolate, had been on the grocer's shelf for decades. The grocery was not simply a gathering of barrels.

Nevertheless, Uneeda's vignette is far less disgusting than other descriptions of the old-time store, including many by people who were not as self-interested as Uneeda. Accounts of late-nineteenth-century groceries sometimes express affection for the potbellied stove and the characters who hung around the store, but rarely have anything good to say about hygiene or the display of goods. Flies, either buzzing around or slowly dying on hanging fly-paper strips, loom large in grocery store memories. Often, they landed in the sticky drippings of the molasses barrel, never to fly again. Sinclair Lewis in *Main Street* described Gopher Prairie's chief grocery: "In the display window, black overripe bananas and lettuce on which a cat was sleeping. Shelves lined with red crepe paper which was now faded and torn and concentrically spotted."

Rural groceries, which stocked hardware, cloth, and other

provisions, were different from urban groceries, which could be found on nearly every city street corner. The rural stores served customers who might not come in for weeks or even months at a time and then bought in quantity. Their customers typically had very little choice about where they could shop, a condition that helped stir resentments to which the giant mail-order merchants that arose during the late nineteenth century responded. They were famous for being dark, cluttered, and none too clean. "The air was thick with an all-embracing odor, an aroma composed of dry herbs and wet dogs, of strong tobacco, green hides and raw humanity," wrote Gerald Carson in his history of the country store. "The storehouse was usually dark and dim, with no windows along the sides. . . . There was a suspicion that the trader liked the twilight so as to discourage exhaustive scrutiny of his goods." Only the storekeeper knew where the stock was, and he wasn't really sure. Carson estimated that such a store would serve about a five-mile radius.

The rural stores served an agricultural economy; people raised animals for meat, grew vegetables, and often made such items as soap themselves from wastes produced on the farm. A country grocer's customers were also his suppliers, bartering their produce for items he stocked. Sometimes those who mistrusted the store's products had very good reason: they knew the goods they traded there were inferior. One common story involved a farmer's wife who found a mouse drowned in the cream and then churned it into butter for trading at the country store. It is less important to know whether this happened often, or at all, than it is to recognize that people feared that it did. "Nearly everything was different from what it was represented," recalled P. T. Barnum of his youth as a store clerk in Bethel, Connecticut. "Our ground coffee was as good as burned peas, beans and corn could make, and our ginger was tolerable, considering the price of corn meal." Folklore and anxiety placed sand in the sugar; water in the rum; dust in the pepper; flour in the ginger; chicory in the coffee; lard in the butter; water, lye, and tobacco juice in the whiskey.

Those who went to city groceries also worried about adulterated goods, but unlike their rural counterparts, they were not among the store's suppliers and thus not party to the adulterating.

93

City groceries served perhaps a block of households in each direction. A given household might make several small purchases there each day, more often picked up by children or servants than by the lady of the house. It was easy to send a child to the store because the range of products was small, there were few competing brands, comparison shopping for price was close to impossible, and there was little to encourage shoppers to purchase on impulse.

Contemporary writers noted that new houses did not need to have large larders because the corner store served that purpose for most urban households. The shrinking of the area allocated for food storage coincided with the decline of the household as a center of production and its redefinition as a center for consumption, with the woman of the house as chief consumer. The advent of central heating also meant that foods were more difficult to keep around the house. Later, better packaging, electric refrigerators, automobiles, and supermarkets would force kitchen storage to expand once again.

Urban groceries also lacked marked prices, and the cost the grocers marked down in their books depended on a number of variables — including the customer's presumed wealth and how long it would probably take the customer to pay, both of which drove the prices up. Goods were often tagged with codes to reflect the storekeeper's cost of the item and the asking price, but many storekeepers were said to have a four-tiered price scale: the price for those who paid cash, for those who had steady jobs and would pay on payday, for those who had to be dunned, and for those who had to be sued. Food expenditures constituted a much higher percentage of people's disposable income than they do today. An estimate made immediately before World War I said that as many as half of all urban households were spending half their income for food.

Still, despite all their differences, city and country stores had much in common. City and rural households alike made many of their own necessities. As we have seen, canned goods were making inroads throughout the century, but still, processed foods accounted for only a tiny portion of a grocer's business. Moreover, advances in jars for home canning had made those foods a more important part of most people's diets, both in the city and the countryside, than were commercial canned goods.

Today's supermarket contains thousands of products whose basic contents are flour, sugar, and some kind of fat. A grocer of the prepackaging era would sell those three commodities and leave it to the buyer to make the finished product. At least in urban areas, buying such ingredients was probably more convenient in the age of the corner store than it is today, because the store was near at hand. Contemporary convenience is defined by a wide choice of products that are processed to avoid labor at home. But the task of going to the store and choosing is more onerous.

The intimate relationship that developed between the customer and the owner of the general store or the corner store was both a strength and a weakness of the system. The strength was that each transaction did not require a separate decision but was part of an ongoing personal association. The problem was that, like all relationships, especially those that involve money, there were opportunities for abuse. And customers were likely to perceive that the merchant held most of the power in this relationship. Nearly everything at a grocery store was purchased blind. Shoppers had no real way of knowing whether this week's flour was the same as last week's. (And until the advent of packaged products and advertisers' dire warnings, they probably didn't really care.)

Typically, owners of corner stores bought from one or more wholesalers, who bought in turn from millers, sugar refiners, spice merchants, rice growers, and other manufacturers of grocery products. Because they dealt in products that were essentially interchangeable, grocers could play one wholesaler against another to get the best price. Similarly, the wholesalers could deal with manufacturers. In theory, at least, this was also a good system for the ultimate customer, because such competition should lower the price at the store. Such free-market theory would require, however, that consumers feel free to shop at different markets to get the best price. In fact, few seem to have done so, perhaps because they depended on credit. It is also important to note that the concept of being a "smart shopper" did not emerge until the rise of the home economics movement at the turn of the century. Buying provisions was a necessity, but few would have thought it a skill. People stayed with their local grocer long after the relationship had acquired many of the hurts and suspicions to be found in a bad marriage. Packaged products were a device by which grocery manufacturers escaped from the competitive commodities market, in which wholesalers could squeeze their profit margins. And one way to do this was to play on customers' inherent mistrust of their local grocer.

The emotions that pass between customers and merchants can easily get in the way of sales. One product on a shelf can alienate a customer without preventing the purchase of another package a few feet away. But if you put both products behind a counter, with an impatient salesclerk standing in front of them, it's likely that neither will be bought. The trend of retailing over the last century and a half has been toward making the goods accessible to the prospective buyer, while gradually eliminating the role of the intermediary.

The crucial first step in this process was the establishment of fixed prices for merchandise, replacing the practice of bargaining. The first advocates of fixed prices were the Quakers. They viewed the issue as one of fairness for all, and it also was consonant with the Quakers' campaign to purge material possessions of their emotional power. But it was also recognized as a good business practice: those who feared they might not be good bargainers felt that they were treated equitably by Quaker merchants. Ultimately, the widespread adoption of fixed uniform prices did not relegate objects to the purely utilitarian status the Quakers believed was appropriate. Rather, it hastened the transition from human transactions to impersonal but emotionally charged encounters with covetable objects.

The widespread application of a fixed-price policy did not come until the rise of the great department stores. Bon Marché in Paris means "good price," and what that meant in practice was prices that were marked and uniform, a system instituted in 1852. The stores that opened in American cities during the next several decades — A. T. Stewart and Macy's in New York, John Wanamaker in Philadelphia, Marshall Field in Chicago — followed this innovation. These stores, many of which looked like outsize Renaissance palazzi, sought an affluent buyer and did provide a large amount of pampering and flattery for their customers. Nevertheless, the price tag was a democratizing instrument, one that stated quite clearly that the only thing that separated you from any particular one of the finer things of life was a precise amount of money. Previously, the finest stores kept their goods out of sight and brought them out only for the most serious customers. But in department stores it was possible for anyone who wandered off the street to take inventory of the many things it was possible to buy.

The next step came in 1879 when F. W. Woolworth opened his first five-cent store in Lancaster, Pennsylvania. Woolworth was the first retailer to define his products not by what they were made of, what they were used for, or who would buy them, but rather by how much they cost. Producing for specific price points is one of the cornerstones of modern retailing, but Woolworth was the first to make consumption shape production, rather than vice versa. Industrial processes were driving down the cost of producing goods. That meant that the number of useful things Woolworth could offer at a low price was rising every year, provided that he could bring the same economies of scale to selling that had already been brought to manufacturing.

From the time he opened his first store, Woolworth envisioned a chain. During the first few years, he capitalized his expansion by enlisting partner-managers to open Woolworth stores in different cities, but in 1888 he laid out a master plan to open about a dozen wholly owned stores a year in all cities east of the Rockies with a population of 50,000 or more, and in many cities of 20,000 to 50,000. There were fifty-five stores by 1900 and, following a merger with several other five-and-ten chains and expansion to Great Britain and Germany, 1,038 stores by Woolworth's death in 1919. The chain quickly began to have the economic clout to make large orders at low prices, establish low profit margins, and make money through volume sales.

The F. W. Woolworth Company's *Fortieth Anniversary Souvenir* gave this account of the company's method:

> [A] buyer had brought to his attention a certain popular finger ring which retailed at 50 cents. He told the manufacturer that he wanted this ring for the Woolworth stores to retail at 10 cents. "Absurd!" was the manufacturer's comment. "I can't make that ring so that you can sell it at 10 cents. Anyway I am selling plenty as it is — more than four hundred and fifty dozen this year." The buyer said he could use a great enough quantity to make the deal profitable for the manufacturers. The rings were produced. Woolworth's sold them at a dime each. The manufacturer made more money than ever before, and the people got genuine gold filled rings at 10 cents — the same in every way as those sold for the higher price. The buyer said he could use a "great quantity." He actually did use 5,000 gross, or 60,000 dozen. Quite a difference from the 450 dozen that the manufacturer thought represented a big business!

Woolworth's store design took the sense of accessibility pioneered by department stores and thoroughly democratized it. Falling prices for plate glass enabled Woolworth to have generous display windows on Main Street and an interior filled with glass cases that allowed the buyer to get a very good look at the objects and their price. Many others were laid out on the counters. In contrast to the old system, in which the clerk stood at a counter with the customer in front and the goods behind, at Woolworth's the customer was in direct contact with the goods. Clerks were available to offer assistance, complete sales, and put purchases in bags. They were specifically ordered not to sell. And store personnel were frequently enjoined by Woolworth himself to spend their time observing their customers' behavior and desires, while refraining from judging them. "Each manager must study the wants of his customers all the time, not try to please his own taste," Woolworth wrote in a memorandum to his managers. "To illustrate: In years gone by, there used to be demand for certain vases, the ugliest ever made, and I was obliged to buy them against my own taste and judgment.

And how they did sell! The same thing applies today. Tastes differ and we must have goods for all."

Woolworth's represented shopping without fear. The range of prices was known. In the earliest years, everything in the store cost five cents. When during the early 1890s Woolworth was forced to add ten-cent items, they were kept in a different part of the store, so it would still be impossible to make a mistake. Loss of foreign suppliers and material shortages during World War I forced Woolworth's competitors to admit items costing as much as a quarter, but Woolworth held the line at nickels and dimes. Customers didn't need to fear embarrassing themselves by asking for something they couldn't afford. The sales force entered the transaction only after the customer had decided independently to make a purchase. It was not as pleasant a personal transaction as a wealthy patron might have at a specialty store where he was well known or as a well-heeled customer might expect at Macy's or Wanamaker's. But most people had never received such privileged treatment. Simply to have a place where there were hundreds of items that they could choose from and where their presence and spending was welcome more than compensated for the loss of the deferential, flattering personal service they had never experienced.

The other great merchandising company that helped establish the idea that the distant big company was the true champion of the little guy was A&P, the Great Atlantic & Pacific Tea Company. This started as a single store in New York, grew to a small chain selling tea and a handful of other high-value products, and then became a huge chain of small groceries. The original premise behind A&P was to take smaller profit margins on items like tea and coffee that were traditionally high-profit items for independent groceries. It also promoted what it called the Club Plan, by which people could get larger discounts by placing large orders along with a group of friends or associates.

Meanwhile, it did not stock even the relatively limited inventory of a standard urban grocery. As late as 1911, sixty-two years after its founding, A&P was carrying only twenty-five different kinds of products, including only a single choice of many staple items. Only after 1913, when A&P began to institute its no-frills, no-credit, no-deliveries, no-trading-stamps, no-premiums economy stores — they closed when the clerk went to lunch — did A&P really begin its rise to become the country's dominant grocer. But once John A. Hartford, the son of A&P's founder and the architect of its expansion, decided to follow the Woolworth model of volume rather than margin, the expansion went very fast. Between 1914 and 1916, A&P opened 7,500 new stores, though only about half showed enough profitability to be kept open. By 1925 there were 14,000 A&Ps.

The speed of this expansion is perhaps a bit less impressive when you consider that most of these A&Ps were small storefronts, carrying a few hundred items. The average investment to open a new A&P was about a thousand dollars. An A&P was about the same size as a typical urban grocery store, though it could be operated more cheaply because of the cash-and-carry policy. Most of the stores had only one employee, who received a commission on sales that exceeded a set minimum amount.

A&Ps were clearly less convenient than their independent competitors. Their advantage was purely a matter of price. And the standard for making the price comparisons was not bulk commodities but packaged, branded products. If buyers knew that a box of Uneeda was the same at the A&P and their local corner store, they were more likely to purchase it where it cost a penny or two less. Whatever relationship they had with their grocer became less important. Even worse for the independents, some shoppers began to patronize the cheaper cash-and-carry stores when they had the cash and return to their independent merchants when they needed credit. The independents' cost of doing business was thus pushed up, increasing A&P's advantage.

This produced a tricky situation for the manufacturers of packaged products. They had only recently been able to convince local grocers to stock their products. Despite the growth of A&P and such other regional chains as National Tea, Kroger, Jewel Tea, American Stores, and Safeway, the manufacturers needed the independent grocers. Products available only at economy stores might be thought cheap. Independents set the standards. Besides, the essence of a national brand is to be available everywhere, not just in a handful of chain stores.

There is also an inherent hostility between a national brand and a national chain. A&P had its own famous brands of tea and coffee and a few other products. The national brands had to defend themselves against the power of any retailer to send them to oblivion. Thus, although the economy stores were made possible by packaged products, some famous brands — such as Cream of Wheat — refused to sell to them because of their cost-cutting policies.

Manufacturers of packaged foods, cosmetics, and most medicines were generally supportive of the first major governmental foray into consumer regulation, the Pure Food and Drug Act of 1906. There had been calls for some sort of regulation of packaged products as early as the 1830s, the era of Swaim's Panacea. But while some European countries became involved in regulation, especially with products that claimed medical benefits, in the United States what people bought and swallowed was always considered to be a private affair.

Two pieces of writing changed people's attitudes. In 1905 *Collier's* magazine carried the first installment of "The Great American Fraud," an exposé of the patent-medicine industry. And in 1906 *The Jungle*, Upton Sinclair's account of the contemporary, industrialized, but far from immaculate slaughterhouse, shocked the nation. Together, these two powerful reformist accounts produced rapid results in Congress.

There is a certain irony in the grocery manufacturers' advocacy of such legislation. They were regularly using selling techniques adapted from patent-medicine hawkers. And more directly, many parts of the rank, bloody animal carcasses that Sinclair described in such detail made their way into packaged products. The focus of *The Jungle* was on meat intended to be sold fresh, but the same Chicago packers were also canners. The conditions Sinclair described were certainly more serious than the danger that a bit of mackerel odor might infect a bag of crackers. A careful reader might conclude that what was inside the package might not be as pristine as the manufacturers of food and cosmetics implied.

Nevertheless, the manufacturers of packaged goods had little choice but to support health regulation, while working behind the scenes to make sure it wasn't onerous. From at least the time of Gail Borden, cleanliness was viewed as one of the chief advantages of packaged products. It is easier to make sure a single large production facility is hygienic than to clean up the entire system of transportation, warehousing, distribution, and sales. A desire for clean food would ultimately work in favor of the manufacturer of packaged goods. Besides, from the time of the first early packages three centuries before, they were festooned with endorsements of different kinds. A government inspection sticker was another form of endorsement, one that enhanced the package's ability to give the consumer peace of mind.

Although the *Collier's* series and *The Jungle* were immediate causes for regulation, it's also likely that campaigns like Uneeda's had already changed expectations about the cleanliness and effectuality of what people bought. A generation earlier, many people were happy to be able to buy anything at all. But now they had been convinced that they ought to have standards. Occasional, grateful purchasers had been turned into regular customers — and they were told that they had a moral obligation to be good consumers.

The great theorist of America's invention of the consumer society was the Illinois-born, German-educated economist Simon N. Patten, whose key book on the subject, *The Consumption of Wealth*, was published in 1889. In this short book, he argued that consumption is a key mechanism of human progress. He argued that issues of consumption come into play after people's physiological appetites have been sated, and people need a reason to keep on living, a novel situation he found in America and much of Europe in his own time.

In a variation of Darwin's theory, he maintained that by

increasing people's choices of what they can consume, society can be improved from the bottom up. "Our large cities ... offer so many choices in consumption that men steadily degenerate or improve. The first class are swept off by disease and vice, and the second class are left to enjoy the advantages of improved consumption." Thus, he argued, although industrial production always seeks to make use of "the cheap man," that creature is constantly being improved as those who make the best consumption choices survive. "A nation with a few intense pleasures might also be compared to a tree with a single tap-root, while a second nation with many equal pleasures would be like an oak with a thousand roots branching in all directions," Patten argued. "The power of the two nations to resist the destructive tendencies forcing them to a lower scale of living would be as different as is the power of the two trees to resist the uprooting force of a fierce wind."

There is no evidence that Frank Woolworth, Henry Crowell, John Hartford, or any of the other consumer pioneers made any merchandising decisions based on what Patten wrote. Patten's distinction may have been that, unlike others who had great insight into the nature of the society that was emerging in his time, he chose to write a book rather than open a store. His work certainly had some business insights that were prescient, such as his prediction that cheap sugar would make sugar-based drinks a larger business than alcohol. At the time, Coca-Cola — invented only three years before — was barely known outside of Atlanta.

But he was most innovative in exploring the psychology of consumption, the essence of which is choice, or at least the illusion of choice:

> It is not the increase of goods for consumption that raises the standard of life, but the mental state of a man after the order of his consumption has been changed so as to allow greater variety. The standard of life is determined not so much by what a man has to enjoy, as by the rapidity with which he tires of any one pleasure. To have a high standard of life means to enjoy a pleasure intensely and to tire of it quickly.

103

As this quotation shows, there was a dark side to Patten's thought that slightly later advocates of progress through consumption rarely echoed. Such spokesmen came largely from the advertising industry, which at the turn of the century had transformed itself into a business concerned with psychology and salesmanship. Advertising's solution to the dilemma of making proper consumption choices was the advertised, branded, packaged product.

W. H. Black, advertising manager of the *Delineator* magazine, put the moral issue this way in a 1907 book, *The Family Income*:

> In all the duties of a good woman's life there is none more sacred than this — the duty of Wise Buying. And one of the most grievous wrongs you could do to your home life or to yourself would be carelessness when exchanging for home needs the money that represents almost the life-blood of one who loves you and has done his best.

Black's assertion that women bear the chief responsibility for the spending of family income was probably accurate. Men were far more likely to be employed outside of the household than women were. This was, nevertheless, a recent phenomenon, and Black's description of the earning male and spending female was not yet commonplace. The problem he saw was that women are innately economical and might hunt for bargains rather than make the more expensive choice of buying branded, packaged items. He argued that this higher-priced choice is usually the more conservative choice, because, "the man . . . who builds up a trademark brings a higher morality into business and a larger safety for all consumers."

During the last two decades of the nineteenth century, family purchasing had changed dramatically. It had previously taken a very high percentage of a household's cash income, but relatively little time. Choices were few, and people were, in general, more concerned about producing their livelihood than purchasing a standard of life. But as industrialization made home production less important, increased cash income, and offered an unprecedented array of choices, consumption became something for people to worry about. Here is how Black expressed the problem:

> Suppose every time, before you bought oatmeal or sugar or thread or soap, you had to examine every bit of your purchase to make sure that the oatmeal and sugar were full weight and free from taint, dirt or adulterant; to make sure that the soap would really do as much work, do it as easily, as well, and as harmlessly as any other soap you could buy for the money; and to make sure that the thread would hold wherever you stitched it, without causing loss and extra work by breaking. Think how all that extra time that you would have to spend on every purchase would add to the cost of articles that you buy every day.

Before packaged products, people would not have considered making such an inquest into their purchases, whatever suspicions they might have harbored. Such an expectation of maximum effectiveness was new. In fact, Black and others like him in the advertising industry were, at that moment, inventing it. By making women's burden of choice so onerous, and the stakes a life-or-death matter, they sought to make shoppers resort to their name-brand, packaged friends.

5 *Serve yourself*

By the mid-1920s, the list of things sold in packages included such unlikely items as casters, towels, inner tubes, bait, nails, queen bees, bacon, belts, lard, hinges, and saws. One brand of wrench had been successfully marketed in a gift box. Manufacturers were beginning to understand that packages didn't merely protect the product and help sell it. They also made new selling opportunities and new products possible.

In their influential 1928 book, *Packages That Sell,* Richard B. Franken, who lectured at New York University on the interesting new topic of the psychology of advertising, and Carroll B. Larabee, an editor of *Printer's Ink,* described the stages of the evolution of packaging. They argued that for the first company to use a package within a particular product category, that act itself represents such a big change that, with rare but significant exceptions, little attention is given to the particular characteristics of the container. Merely offering a package is enough to make the product stand out. But while the decision to go into a package may have been a good one, the authors argued, the company has only a short-term edge before other companies learn from its success:

The competitor, however, has an advantage. He sees that just getting a package will not be enough; he must get a better package. . . . The new container not only fills an economic need, it fills a sales need. . . . It adds to the appearance of the store and helps the clerk to sell. In other words, it not only contains, it *advertises.*

The original packaging innovator, along with the other companies in the industry, has little choice but to respond. At that point, the product category enters what the authors termed the "competitive stage":

Therefore they build "convenience packages" equipped with "shaker" tops, "captive" caps and "spreader" tips. The entire evolution of such a product as shaving soap from mug through stick and powder to cream is also an evolution in package construction.

This competitive stage was, in the minds of these authors, the normal state of affairs in a marketplace, an economy, a society, in which the ability to influence what people buy was becoming the single largest factor in successful business.

The relationship between packaging and advertising seems clear and fundamental: advertising creates a widespread awareness of a product, and packaging completes the message, closes the sale, and in many cases continues to sell while the product is in use. We have seen how a few pioneers, such as Crowell and Mennen, built large and enduring businesses by integrating packaging, advertising, and promotion. In Mennen's case, innovations in both package design and engineering protected the products against refilling and thus provided a level of reassurance that is particularly important — and salable — for products applied directly to the body. But such a unified understanding has been comparatively rare, even up to the present day. Advertising has always involved more money, more talent, more glamour. While packaging remained tied to the manufacturing process and to the product itself, advertising suffused the environment — on posters, the sides of barns, and the interiors of streetcars, in magazines and newspapers and later on radio. Because it was separate from the product, it became its own area of expertise and was dominated from the turn of the century onward by advertising agencies.

Packages were close to the product, and thus they were designed by the owner of the company, or his wife or nephew, or were left to the production department. Some of the world's best-known, and even best-designed, packages emerged from such a process. Wrigley's Spearmint gum, with its green-arrow packet, was introduced in 1905, and rarely has an idea about taste, about ingredients, and about the emotional reward offered by a product been more economically and powerfully expressed. It is one of those designs that has become so familiar that one is rarely conscious of it, but the strong, recognizable graphic still does the job. In 1993 *Financial World* estimated the value of the Wrigley brand at more than $1.6 billion. It's hard to say how much of that can be attributed to the package with the arrow, but it's probably quite a lot.

Perhaps more typical, though, was the Uneeda biscuit. Its patented, two-layered package represented a real breakthrough in keeping crackers fresher. Its memorable name was dreamed up by the N. W. Ayer advertising agency, whose success in dramatizing the existence of the new package was discussed earlier. Its advertising also included one of the great advertising icons of the early twentieth century, the slicker-clad boy carrying his box of Uneeda through a downpour. But the box itself, with its distinctly Victorian purple-and-white frame, was inoffensive at best. It identified the product but did nothing to express the freshness either of its crackers or of the marketing strategy. The box did appear in the advertising, but only in the advertising were the attributes of either the package or the product dramatized. It was not atypical. According to Franken and Larabee, in 1900 only 7 percent of print advertising contained pictures of the package.

In fact, most manufacturers were a lot less interested in their packages than was Uneeda. The result was a mismatch between the up-to-date literary and artistic orientation of those who worked for advertising agencies, and the often high-Victorian sensibilities of the managers and manufacturing specialists who had control over most packaging.

During the first three decades of the century, however, a convergence between packaging and advertising began to be evident. One major reason was what might be termed the Aunt Jemima Effect. Many manufacturers were learning the lesson first demonstrated by the pioneering pancake ready-mix — that packaging makes it possible to combine familiar ingredients to create novel products. Bottled salad dressing, an unknown product in 1916, became a $17.5-million business by 1922. As the patent-medicine makers had discovered centuries earlier, a proprietary mix is particularly valuable because once loyalties are established, they are not easily broken by a few cents' difference in price. And although consumers were still resisting packages of such simple commodities as salt and sugar, they were willing to try products that took such ingredients and made them into something more convenient. While commodity packaging was, for the most part, designed to create a feeling of confidence, the packaging for these novel convenience products had to help potential buyers understand the result. "Not only the package as a whole," wrote Franken and Larabee, "but every single part of it, is so important that its consideration cannot be neglected without putting an obstacle in the way of the product's sale."

What was happening was a subtle but profound change in the character of production and consumption. The advertising industry had begun to help people sell what they made, and packages served for shipping, protection, and brand identification. But it soon became apparent that packaging allowed the creation and marketing of new kinds of products, and advertising had the power to make these new products as much a part of people's lives as more familiar, "natural" commodities. It seemed clear that advertising and packaging for such products had to be considered together.

One of the strongest traditional distinctions between packaging and advertising was the idea that advertising was everywhere, while packaging existed at only a few points of sale and was not always visible. This boundary was eroded when it was discovered that the right package can colonize hitherto undeveloped territory and increase the number of points of sale.

The classic case of this phenomenon involved Lifesavers, the ring-shaped hard candies. This started out as a packaging disaster. Lifesavers were originally packed in a glued cardboard box. The problem was that as they sat on the shelf, the taste of the glue migrated to the candies and made them inedible. Meanwhile, the mint taste migrated to the cardboard, where it did no good. Faced with a refusal of grocers to stock the candy, the founders were able to sell the company to a group of partners who were apparently unaware of the problem. In fact, the technical solution was quite easy. The candies were wrapped in metal foil, which retained their flavor and kept foreign flavors out. The package could also be easily resealed by the purchaser. The cardboard box, which tended to hide the distinctive shape of the candies, was discarded, and a paper collar was substituted.

The marketing challenge was much trickier, because the logical sales outlets were already alienated and saw no good reason to give Lifesavers a second chance. Out of desperation, the company decided to try selling the candies in saloons, and they designed displays to go next to cash registers. The suggestion was that a bit of mint on the breath might make it less obvious where the man had been spending his time. Enough saloon owners were willing to give up a foot or two on the bar to make some extra money, and it worked. The breakthrough came in 1915, when the large United Cigar Store chain installed Lifesavers displays on counters of hundreds of its outlets, with a large sign that read 5 CENTS. Lifesavers candy seems to have been the first product to prosper entirely as an impulse purchase, bought by those whose primary interest was in something else entirely. It colonized the frontiers of selling. The short cylinder shape of the Lifesavers package enhanced its incidental character; it was something for the pocket, like coins or keys, rather than a box of candy. In 1919 Prohibition closed the saloons that launched Lifesavers, but by that time, the candy was next to cash registers everywhere. The combination of package, place, and promotion helped create a seamless environment for desires to be created and conveniently fulfilled.

At the same time, the advertising industry — both the large agencies and such emerging communications empires as Curtis Publishing Company — was developing the business of market research. Surveys were done of stores, businesses, and countless individual shoppers to ascertain preferences. Curtis was particularly aggressive in moving beyond what people said they wanted and used and determining what they actually bought. In one survey, Curtis inventoried the cupboard shelves of 3,123 households in eighty-five neighborhoods in sixteen states. In another, it collected four weeks' worth of trash from 56 Philadelphia households from different income groups and made its analysis based on the six thousand discarded packages. This 1926 study recognized that garbage is a sincere, and almost involuntary, expression of values.

Psychological factors, which a few of the packaging pioneers had understood instinctively, were also beginning to receive systematic attention. One of the most important of these concerned the sizing of packages. Companies had learned that buyers had strong but unexpressed ideas about the most convenient sizes for specific kinds of products. Franken and Larabee noted that it was well known that people would rather pay ten cents for four ounces of baking soda than twenty-five cents for a full pound. The ability to determine exactly how people balanced the amount they wanted to pay for a thing with how much they wanted to have could contribute enormously to profits. It was obviously a factor in decision making about packaging, but to the extent that such expertise existed, it tended to reside in advertising agencies.

During the 1920s, studies began to be done about the psychological impact of color. The familiar insights — dark colors look heavy, bright colors light, yellow cheap, and small packages are expensive — were emerging from such studies. Such an aura of science meant even the decoration of the box was too important to be left to the boss's wife.

"In a decade trade-marked B.V.D.s' advertising changed the underwear habits of several nations," wrote Clayton Lindsay Smith in *The History of Trade Marks*, published in 1923. "Few men will own that advertising can force them to change their underwear. This seems to show that few men understand the vulnerability of their minds."

*　　*　　*

The combination of packaging, advertising, and research allowed businesses to shift from selling what they made to making what would sell. Packages made this shift possible. Yet to those in the advertising industry, packages too often seemed backward.

"One can company manufactured enough cans last year to reach six times around the world," wrote the advertising visionary Ernest Elmo Calkins in 1920. "Just think of the wasted advertising space on the outsides of these cans. Just think of the ugly, uninviting, tasteless cans of tomatoes, soups, fish and fruits that are sold over the counter each year. If their contents were as insipid as their outside treatment, they would never sell." Calkins, founder of the advertising agency Calkins and Holden, was a key figure in transforming advertising from essentially a writer's medium to a medium of visual communication, combining graphics, illustration, photography, and copy into a unified expression. Although his own tastes seem to have run toward the picturesque and genteel, he hired artists and illustrators who worked in modern styles. "Advertising is the supreme flowering of sophisticated civilization," he wrote. He seems to have been offended not just that ugly cans and boxes were messing up his ads, but that a field so closely identified with his own should fail to recognize the cultural advance advertising represented.

"Why shouldn't a vegetable can look as good as its contents?" Calkins wrote. "Why shouldn't it be its best advertisement? Why?" Calkins launched a package design studio within his firm, and he hired Egmont Arens to run it. And if your idea of sophisticated civilization encompasses such Moderne artifacts as Fred Astaire–Ginger Rogers movies or Chrysler Airflow automobiles, Arens did, indeed, civilize the pantry shelf and the icebox. The red bag of Eight O'Clock coffee with its elegant lettering (and its even more elegant, more expensive sibling, the black-bagged Bokar) were as close to elegance as most Depression-era Americans ever got. Calkins's rhetoric sounds like self-parody now, but there is, nonetheless, some truth in it.

*　　*　　*

Nevertheless, though the arguments of Calkins, Franken and Larabee, and other 1920s theorists of salesmanship still seem compelling, packaging never did become the province of advertising agencies. That doesn't mean that Calkins and his allies went unheeded. Indeed, the advertising function of packaging was widely recognized during the teens and twenties throughout the world, and packaging in general became more visually interesting and closer to other decorative and artistic trends of the time. What happened was that the profession of salesmanship grew beyond advertising and helped reshape everything that was being made, marketed, and used. An important manifestation of this was the rise of industrial design.

The seemingly sudden appearance toward the end of the 1920s of a handful of glamorous form givers, whose imaginations shaped countless products and the popular idea of progress itself, is a frequently recounted story. Indeed, the first practitioners were also among the most assiduous mythmakers. Walter Dorwin Teague, Norman Bel Geddes, Henry Dreyfuss, Donald Deskey, and above all Raymond Loewy became celebrities during the 1930s, largely because they promised to bring an excitement to the objects of everyday life that would encourage people to consume their way out of the Depression. But the field was actually invented during the boom years of the 1920s, in response to some of the same forces that have already been discussed in terms of packaging. The imperative to make what sells did not only affect the container for the product, but also the product itself. The first generation of industrial designers came from backgrounds in advertising illustration, retail display, and theater design. They were communicators, not engineers. Their job was to make products understandable and exciting. In contrast to the European functionalists, mostly architects, who were their contemporaries, they understood that their primary goal was to move the product. And if that meant making a stationary object like a refrigerator appear as fast and aggressive as a locomotive, then so be it.

Industrial design *was* packaging, at least in the broad sense of the term. It was an external expression that protected the product and made it attractive, exciting, and good to have around. Raymond

113

Loewy's famed redesign of the Gestetner duplication machine did not have any impact on the way the machine worked, mechanically. It dealt only with the way people felt about the machine. The strategy was to design a case that would place nearly all the mechanism out of sight and leave visible only what the user had to deal with.

Though this approach has often been derided as mere styling, the idea of designing a machine from the point of view of the user rather than its production was still quite new. Today, most people spend much of their lives using devices they do not begin to understand, and ever so vaguely fear. As the age of the computer has taught us anew, user interfaces are important. It can be argued that Loewy, by hiding the motor of his Coldspot refrigerator, and putting it underneath the food compartment where its heat could only go upward to warm the food, set home refrigerator design back by decades. But he also provided a refrigerator that people felt good about having in their kitchens. He packaged a noisy, power-hungry contraption as a clean, dynamic icon of modernity.

The approach of Loewy and his colleagues of the first generation of industrial designers was, in a literal sense, superficial. It affected none of the things that made the machine work — except the user. But from the user's point of view, the change was profound. The machine had been transformed from a forbidding challenge into a modern convenience. Form followed fashion. Form kindled desire. Form embodied progress. And indeed, form followed function — so long as you accept the premise that the function of a product is to be sold.

Some of the same insights were apparent in another important 1920s innovation, General Motors' introduction of a hierarchy of cars, ranging from the Chevrolet to the Cadillac, each of which would have an annual model change. These linked practices transformed automobiles into markers for two distinct types of progress. One could trace one's personal progress up the economic and social ladder through the move from Chevy to Olds to Buick and perhaps even to Cadillac. At the same time, all the cars were embodying an overall sense of change and improvement from one year to the next, which prodded even those who had made it to the top to buy new in order to keep up. In a country that tends to deny the existence of class, and requires most people to assert their claims to a high place in the society, automobiles allowed people to package their achievements and aspirations for inspection by others.

What happened in the 1920s, in the United States above all, was that advertising values spread beyond the advertising industry itself and thoroughly reshaped the goods that were produced and the ways in which they were understood and used. In effect, every product had turned into an advertisement for itself.

It's not surprising that many of those involved in this broader sense of packaging should also have involved themselves in the design of boxes, cans, and other forms of packaging. Loewy, Teague, Deskey, and others launched packaging departments within their offices, some of which are still leaders in the field. Although they were not engineers, industrial designers were at least experienced in communicating with those who solved the problems of actually making a product or a package. Most advertising men were not. Moreover, the expected life span of a particular package was, in most cases, far longer than that of an advertising campaign. Gearing up to produce a new package involved high fixed expenses that needed to be amortized over a long period of time. The package became, in many cases, the expression of the product's identity and wasn't to be trifled with. People expected their Buicks to look different each year, but not their bags of flour or bars of soap. Besides, changing the package was an awful nuisance, which generated new advertising costs as well. Thus, the challenge of creating packaging that sells proved to be more closely related to the skills of those who sold by shaping products rather than those who sold by creating advertisements.

Nevertheless, industrial designers did not dominate the field, either. Graphic artists such as Jim Nash, Lester Beall, and Paul Rand created very memorable packages, and most package designers today tend to have a graphic design background. Firms such as Landor Associates and Lippincott & Margulies broadened their specialization in package design to embrace corporate identity — the packaging of a company in a way that is consistent, expresses its

115

values, attitude, or culture, and is appealing to both consumers and potential stockholders. In practice, even within industrial design firms and other multidisciplinary organizations, packaging is usually separated from other design specialties because it has its own sorts of demands and expertise.

The period between the wars brought not only the rise of the sophisticated salesman, but another sort of new creature — the professional manager. These were powerful businessmen who weren't the kinds of proprietors who gave birth to the first generation of packaged products. They weren't owners, nor did they necessarily identify themselves with one industry or another. Nor were they the kind of industrial engineers who fitted human behavior to the needs of the machine. Making a large corporation work involved more than smooth-running production lines, and the new specialists sought to integrate finance, personnel, production, and marketing into a single system. The creation of the first business schools in the late nineteenth century was a recognition of the complexities of running businesses on a continental scale. Even though individualists like Henry Ford were the business heroes of the early twentieth century, professional managers were making important inroads. The crash of 1929 left few untarnished heroes. Like industrial designers, the disciplined, professional colonels of industry were repositories for hope that the economy would revive, but in a new shape.

It began to be apparent during the 1920s that packaging was not only a major industry in itself, consisting of producers of boxes, bottles, and cans along with the producers of the materials from which they were made, but also a component of nearly every other industry. While most of the innovations in packaging took place in the food, cosmetics, and retail-pharmaceuticals industries, there was hardly any product that was not either packaged or at least packed for shipment. (And after the lifting of Prohibition in 1933, the liquor industry, another historic generator of important packages, was back in business in the United States.)

Thus, while packaging was the principal concern of very few, it was among the concerns of nearly everybody. What kind of package is used affects how much the product costs to produce and how much can be charged for it. A manager must also worry about the availability of the package and its compatibility with filling machinery and other parts of the production process. Moreover, packaging influences other decisions that affect the bottom line. It has a direct impact on shipping methods and costs and is a tool of inventory control. The appearance and usefulness of the package serve, according to this analysis, to allow the manufacturer to maintain high profit margins in the face of low-priced competition.

In 1930 Procter & Gamble created a new kind of executive, the brand manager, whose role was to coordinate all of the decisions involved in producing and marketing a specific product. This arrangement was copied in succeeding decades by nearly all American companies and many overseas. The institution of the brand manager grew out of the Procter & Gamble tradition of strongly differentiated, often competing brands and packages. The growth of this concept, however, was the result of the growth and diversification of companies and the move of industry from a generation of visionary proprietors to professional managers. Brand managers do not rank high on the corporate hierarchy, but they play a key role in design decisions. There is a strong case to be made that the rechanneling of design decisions to this middle level of the corporate hierarchy stunted the industrial design profession and gave rise to products whose character was cautious rather than imaginative. One thing that is certain is that packages were no longer perceived as the signature of the owner of the company. Rather, their design was a way for the brand manager to make a mark.

In 1931 the American Management Association held its first national conference on what it termed "the rationalization of packaging," and it later launched an awards program. Ever since, it has regularly issued publications on packaging that constitute much of the professional literature.

It is surprising to read in the proceedings of the first conferences that it was still necessary in the 1930s to curb companies' pride of proprietorship in packages. "Consumers were more interested in products which were packaged to fit into their own scheme of things than in those products which picture the largest factory in

117

the world," marketing consultant and package designer Ben Nash said at a 1935 conference reporting on some recent market research. "Yet many manufacturers were so close to their own business that they could not see this fact."

It is also evident that the organization's rationalizing vision left plenty of room for the irrationality of the buyer. "If I were in the tea business," said Clyde Eddy, manager of the merchandising division of E. R. Squibb & Sons, at that same conference, "I would make the package sing of the tea leaves rustling in the breezes of the Indian Ocean. I don't know whether they do it or not, but I would make them do it." The purpose of the package, he said, is to make the purchaser think, "Ordinarily you would pay $5 a pound for tea like this."

In this case, the managerial approach to packaging doesn't seem too different from that of admen like Calkins, Franken, and Larabee. Over the long term, though, the trend toward defining packaging as one piece of a manager's concern, rather than as a high-level corporate decision, very likely had a negative impact on the character and creativity of packaging.

One of the most insightful and widely read critiques of the triumph of commercial values and of the enlistment of art and literature to create a new, packaged world was Sinclair Lewis's *Babbitt*, published in 1922. In the first few pages, Lewis shows George Babbitt as a committed brand-name buyer, a real estate operator in an up-and-coming midwestern city, awakening to the ring of a high-quality, nationally advertised alarm clock, fretting over the purchase of the wrong brand of toothpaste, snatching his tube of shaving cream, and wishing that there were a safe place to put used razor blades. (Gillette obliged a few years later with its dispenser that contained a used-blade compartment.) Then to his underwear: "He never put on his B.V.D.s without thanking the God of Progress that he didn't wear tight, long, old-fashioned undergarments. . . ."

George Babbitt is, of course, a figure of satire, the epitome of the influenced man. (He may even have gotten his name from one of the pioneers of packaging, Benjamin Babbitt, who was the first person to take chips wasted in the soap-making business and box them as laundry powder and is sometimes credited with inventing the pictorial advertisement and billboard.) Lewis's character is empty of meaning or morality, and he attempts to fill the void in his own personality through buying the products that tell the world and himself how far he has come from his hick-town roots. "These standard advertised wares — toothpastes, socks, tires, cameras, instantaneous hot-water heaters — were his symbols and proofs of excellence; at first the signs and then the substitutes for joy and passion and wisdom." Babbitt is a classic American figure, proud of his shrewdness, deeply gullible, making of his life an impressive package whose purpose is to fool himself.

And yet it must be admitted that those BVDs *were* more practical and more comfortable than what had come before and that the advertised brands often had their advantages. You might criticize the society that produces better underwear, but you probably don't want to give up the underwear. This is the chronic dilemma of the critic of consumer culture: those who get read tend to be quite comfortable, and they are wholly implicated in the system. In *Babbitt*, Lewis deals with this ambivalence through the character of the radical lawyer Seneca Doane. "Standardization is excellent *per se*," he tells a European visitor. "When I buy an Ingersoll watch or a Ford, I get a better tool for less money, and I know precisely what I am getting, and that leaves me more time and energy to be an individual in." None of the post–World War I prophets of branding, packaging, and advertising could have said it better. Indeed, this is an idea that appears in some contemporary sales campaigns. "No, what I fight . . . is standardization of thought," Doane adds. But the suspicion remains that predictable goods and equally predictable minds are part of the same package.

Even Franken and Larabee, whose landmark book *Packages That Sell* was mostly the sort of peppy, how-to literature George Babbitt might have admired, nevertheless felt it necessary to address public concerns about packaging. Unlike Lewis, they were not concerned about the relationship between packaging and personality.

Rather, the criticism they sought to quell concerned what is still a hot issue, the apparent wastefulness of packaging. They enumerated forty advantages that packaging affords the producer, the middleman, the retailer, and the consumer. A few of them seem a bit disingenuous now ("Packages may be used when empty for other purposes"), and others seem a bit circular in their reasoning ("Packages add personal satisfaction and pride in their purchase, and prestige to the household").

On the whole, the arguments were much the same ones that could be made today with equal force and validity, such as the reduction of waste through spilling or spoiling and the assurance to the consumer of uniformity from purchase to purchase. The advantages for manufacturers in being able to better plan their production and distribution and offer multiple product lines were perhaps the most compelling, particularly if readers thought of themselves not only as consumers, but as workers. Franken and Larabee discussed the Hills Brothers Company, which employed two hundred workers full-time and added nine hundred for four months a year. By adding several products to its line, something it could not have done without the brand identity created by the longtime sale of a well-known packaged product, it was able to move to a stable, year-round workforce. They also mentioned Procter & Gamble, which in 1924 for the first time guaranteed forty-eight weeks of continuous work to its employees, a stability made possible by advertising and packaging. Year-round production and continuous employment have been, at least until recently, universal expectations, and most people rarely question how they came about. Predictable packages filled with predictable products might well produce predictable lives. But they also helped produce predictable incomes for many families. Despite the philistinism of 1920s booster rhetoric, that fact is hard to laugh off.

In *Babbitt*, the newspaper poet Chum Frink writes a rhyming rhapsody to the comforts of traveling in America, where you can count on encountering the same magazines, the same cigarettes, and the same kind of people wherever you go:

And when I saw that jolly bunch come waltzing in for eats at
lunch, and squaring up in natty duds to platters large of French
Fried spuds, why then I'd stand right up and bawl, "I've never left
my home at all."

Lewis was being satirical, but the sentiment was nonetheless
true, and there were people who realized that there was a dollar to
be made from it. Why not offer the same French fries on the same
platter in the same setting for the same price in numerous locations?
That way, customers would have the confidence to enter your eatery,
and each location would market all of the others. A restaurant could,
for the first time, become a brand, and the most reliable patron was
as likely to be someone just passing through as someone who lived
around the corner.

At the outset, such an approach seemed most suited to the
places in which strangers were most likely to come together, big
cities. Horn & Hardart Automats, the first of which opened in Philadelphia in 1908, brought the promise of the Uneeda biscuit to
food service. It gave an industrial veneer, and with it the promise of
cleanliness and efficiency, to the always chancy endeavor of serving
meals. The implicit promise was that the food was entirely standardized. It is true that the patron who had some time might look from
window to window to compare a plate of meatloaf with its neighbor,
but the goal was to make such comparisons not worthwhile. One
couldn't say that the food was untouched by human hands; you
could get a glimpse of hands behind the scenes placing new dishes
into empty compartments. But the entire experience had a degree
of impersonality that was reassuring. You were not being served by a
possibly unreliable individual but by a trustworthy organization.

The concept of food service as a packaged experience was
extended by White Castle, founded in Wichita in 1921, a year before
Babbitt was published. While Automats and cafeterias were found in
all sorts of buildings, White Castle was found in only one rather
unlikely sort of building. The earliest White Castles were just ten by
fifteen feet in area and seated only three customers. But they were
nevertheless unmistakable, white with a roofline that suggested bat-
tlements and a turret. The name of the business was a description of

121

the building in which its outlets were contained. The whiteness was found both outside and inside. White demands to be kept clean and, in effect, sets a standard for those operating the restaurant. Patrons, whose seats looked directly at the food preparation area, could judge for themselves whether the hygiene was acceptable. Uniforms on the employees also provided the endorsement of a larger organization that implicitly stood behind them. Standardized menus reduced the necessity for making new choices each time you entered a different restaurant.

The whiteness on the inside was not as innovative as the whiteness on the outside. The buildings declared themselves on the street just as boxes and bottles fight for recognition on the counter. They promised brand names and standardized products, and the promise of safety was literally built into the restaurants' design. White Castles were more sophisticated in applying these principles than were most of the packages then on grocers' shelves. Especially after the chain went to a prefabricated building with porcelain enamel-paneled walls, the White Castle buildings became exactly like one another and very unlike their neighbors. The character of cities is defined by the aggregation of its buildings. But buildings that are packages, in seeking to be islands of familiarity in an unfamiliar landscape, gain their power from being alien. Like all packages, they are placeless, and, as with the Uneeda biscuit box, effectiveness stems from their imperviousness to local conditions.

"When you sit in a White Castle," said a 1932 company brochure, "remember that you are one of thousands; you are sitting on the same kind of stool; you are being served on the same kind of counter; the coffee you drink is made in accordance with a certain formula; the hamburger you eat is prepared in exactly the same way over a gas flame of the same intensity . . ." The litany of sameness went on and on, defining a vision of democracy grabbing a bite to eat. No man's hamburger is better than any other — and no worse either.

Several other chain restaurants, including the blatantly imitative White Tower chain on the East Coast, followed the White Castle pattern of small, distinctive urban outlets. Howard Johnson's, founded in 1925 as an ice-cream stand in suburban Boston, began in the mid-1930s to colonize the highways newly filling with drivers. The highway landscape had already produced an array of quirky buildings, whose unusual shapes were designed to catch the eye of the driver. Howard Johnson's restaurants used some of these tricks. Their orange-tiled roofs offered high visibility, and the company chose sites that were visible from a great distance. And the company appropriated the fairy-tale imagery of the illustrator Maxfield Parrish to make the experience of stopping at the restaurants both playful and mythic. Some of their architectural details were domestic. Their fantasy image contrasted strongly with the look of modernity that many of the urban chains sought. But unlike other highway architecture, theirs was not an improvised embodiment of personal idiosyncrasies. Howard Johnson's restaurants were substantial and even institutional in their appearance. They presented reassuring architecture for the middle of nowhere, refuges in the unfamiliar territory between where you had been and where you were going.

The other key creators of packaged environments during the pre–World War II era were the oil companies. The breakup of the Standard Oil Company in 1911 resulted in the creation of five enormous, competitive companies, each of which was involved in petroleum from the ground to its final use. In addition several other major oil companies, such as Texaco, Shell, and Gulf, were interested in serving the new automobile market. Like the producers of other products, they wanted to build consumer loyalty and be able to predict and protect their market share. But although distinctive kerosene containers had once been a factor in competition among oil companies, by the time the automobile became widespread, gasoline was pumped directly from storage into the customer's automobile, without any container.

The challenge for the oil companies was to brand their products without using a package. The first solution was a distinctive logo to be used on signs. Shell's bright yellow scallop shell, Texaco's red star, Socony's flying red horse, and other oil company symbols began to appear throughout the world. At first, pumping gas was merely an adjunct to other businesses, but because oil company signs had to catch the attention of someone in a moving automobile, they soon took on a dominating scale. At the same time, the oil companies were concerned about public perceptions that filling stations were dangerous and dirty. The oil companies did not own the vast majority of their stations, but they were able to use design as a way to protect their image and, to some degree, control the behavior of their retailers. It was to the dealers' advantage that their stations resemble those clean, pleasant places depicted in the oil companies' advertising. During the 1920s, many stations sought a civilized appearance. Many Gulf stations were small, hip-roofed classical pavilions, while Pure stations took on the appearance of picturesque English cottages. The scale of the stations changed during the 1930s, as filling stations became full-scale service stations, complete with uniformed attendants. Standardization became even more important as the oil companies sought to identify themselves with science and modernity.

In 1937 Texaco introduced its new Type A station, designed by Walter Dorwin Teague Associates. This station was faced in bright white panels, similar to those used in White Castles. A cleaner gas station image was particularly important for women, an increasing portion of the driving, as well as backseat-driving, population. The new design was introduced along with the concept of "Registered Rest Rooms," whose intent was also to make women less frightened of the stations. The overall image sought was clean efficiency. The only ornament on the station, other than lettering over the service bays that said MARFAK LUBRICATION and WASHING, were red stars and a ribbon of three parallel lines that ran around the top of the station. In some stations, this ribbon came off the wall to define a canopy over the pumps. It was primarily the work of Robert Harper, a designer who had earlier worked for Loewy and headed the redesign of the Coldspot refrigerator. Like many of the best packages, this design was distinguished for communicating its message clearly, using absolutely minimal means and not including anything to cloud the message. It was a great success, and by 1940 more than five hundred Texaco stations had adopted the design. Some still exist.

In 1933 a Du Pont subsidiary sent a newspaper to Depression-buffeted retailers. Its banner headline read "Prices

Raised, Sales Increase.'' Such behavior is probably the dream of all store owners, but it is rarely found in real life. Nevertheless, Du Pont had sales figures, and even some controlled marketing experiments, to back up the assertion. It could not happen, however, without a miracle ingredient, one that was, not surprisingly, manufactured by Du Pont.

The secret was cellophane, a wood-based product that was a sort of chemical cousin to rayon. It had been invented in 1911 by a Swiss chemist in an unsuccessful effort to find a coating that would keep tablecloths from becoming stained, and it was first manufactured in France in 1913. Du Pont licensed that patent in 1923 and began producing it the following year. In 1927 Du Pont made a crucial improvement of the material when it developed moisture-proof cellophane, allowing the material to be used as a protective covering for food. This clear, flimsy product remained close to invisible, however, until a few years later when Du Pont began a pioneering marketing campaign to make what had been a little-known component into a household word. The company's success in creating a celebrity substance was affirmed when, in his 1934 song "You're the Top," Cole Porter climaxed a list of superlatives with the accolade "You're cellophane."

In a sense, the popularization of cellophane was analogous to the introduction of the machine-made paper bag more than sixty years before. Like the paper bag, cellophane was often a competitor to packaging because it allowed individual merchants to pack unbranded products in an attractive way. Du Pont's promotions showed retail displays featuring cellophane containers filled with macaroni, nuts, beans, spices, vegetables, and countless other commodities. Many of these products had long been marketed in distinctive packages.

Still, cellophane's challenge to packaged brands was relatively short-lived. The promotion of cellophane benefited packaged brands in several ways. It immediately gave manufacturers more options in designing attractive packages with windows so that shoppers could see what they were buying. Cellophane launched the era of packages wrapped in clear plastic films. These clear coverings, engineered to provide the kind of protection a product needs, today fulfill many of the protective functions people expect in a package.

Moreover, by increasing the number of items that could be sold without any sales assistance, cellophane wrapping encouraged the growth of self-service stores, the ideal environment for packaged items. Like paper bags, cellophane wrappings reduced the friction of salesmanship and accelerated purchasing and consumption. And, most dramatically, this transparent material proved that there is really no such thing as a neutral container.

The obvious advantage of cellophane is that it makes products visible. What's only slightly less obvious, but which was a major selling point for Du Pont, is that the reflectivity of the cellophane gives many products a sparkle that they do not really possess. If questioned, the shopper would know that this sparkle was the cellophane and not the handkerchiefs or kidney beans the cellophane package contained. Nevertheless, there is every reason to believe that the glistening cellophane contributed strongly to the perception that the goods inside were fresher, cleaner, safer than those packed loose, or even than goods packed in conventional opaque containers. One Du Pont promotion to retailers spoke of a department store that dealt with the problem of slow turnover by wrapping all its current stock in cellophane. The wrapping refreshed the tired products, which sold out quickly.

There were certain kinds of products for which cellophane worked wonders. In many cases these were the sorts of things that offer immediate temptation for the shopper, but do not generally appear on a shopping list. For example, despite a generation of promotion by Nabisco and others of the advantages of purchasing cookies in sealed packages, loose cookies remained competitive because they whetted shoppers' appetites. Cellophane promised the sanitary advantages of the wrapped cookie, along with the direct appetite appeal of loose ones. The advantage of cellophane for marshmallows was even more dramatic. The allure of marshmallows is irrational, not to say infantile. Marshmallows are not an idea but a craving. They are not something you write on a shopping list, but if you see them you might grab them. Yet, unlike cookies, they cannot be sold loose, because there are few things less appetizing than a dusty marshmallow. Thus, it's not surprising that when

126

marshmallows went into cellophane-wrapped packages, sales increased tenfold.

Shoppers' often false assumption that cellophane-wrapped products were untouched by human hands made it successful for products intended for babies and young children, while its success in wrapping hardware products that do not need any protection at all was more difficult to explain.

Even fruits and vegetables, many of them wrapped in their own natural packages, were found to benefit from the sparkle and implied cleanliness of cellophane. "The 'Elberta' has a peach of a wrap," declared a 1932 Du Pont ad that appeared in *Good Housekeeping*, "but Nature's best peaches are outclassed by Cellophane."

The visual alchemy of the gleaming, clear wrapping had the greatest impact on meat. Long after the invasion of the grocery store by branded, packaged products, the butcher remained a craftsman. The butcher depended on the confidence of his customers, and this depended, in turn, on the way in which the butcher handled the meat, carefully trimming and dressing it as the buyer watched. It was not really practical for buyers to inspect the meat, at least until they got it home. Each meat purchase was an incident in an ongoing relationship, and a pleasing personality was a very important trait for a butcher. Cellophane changed all that. It became possible for shoppers to inspect cuts of meat closely at the same time it allowed cuts of meat to be presented in a way that looked fresh and pristine, without any reminders of the violence of being a carnivore. (Vinyl film, introduced in 1940 by Dow Chemical under the trade name Saran, brought even more luster and eye appeal to the meat department.)

The butchering of meat had long been an industrialized process — almost, but not quite, to the point of sale. The methods, invented in the great Chicago packinghouses, for systematically breaking a carcass into its many salable components became, during the late nineteenth century, a widely known model for continuous-flow production. The meat processors' disassembly line was the mirror image of Henry Ford's automobile assembly line and has often been cited as an inspiration for it. But this highly rationalized and mechanized process ended in a personal transaction between a

127

housewife and a genial fellow wearing a bloodstained apron and wielding a big knife. His job was the final trim that would provide a cut of meat perfectly appropriate for her needs.

Extending the industrialization of butchering all the way to the retail level through the cellophane wrapping of meat removed this element of individual responsiveness. Instead, it made it possible to greatly expand the size of meat-retailing operations and thus to offer such a wide choice of sizes and cuts that the shopper would not feel deprived. Indeed, such sales techniques gave shoppers a sense of greater control over their purchases. They did not have to depend on their butchers. They could see what they were being offered and make their own judgment about what to buy.

The logical outcome of this thrust toward large outlets for precut, cellophane-wrapped meat is, of course, the supermarket. The open-topped refrigerated cases full of precut meat helps define the supermarket; there had never been anything quite like it before.

It would be absurd, however, to argue that cellophane and its role in replacing the personal-service butcher was itself responsible for the rise of the supermarket. Cellophane was but one of many innovations introduced during the first four decades of the century that helped to make consumption as rapid, predictable, and impersonal as factory production processes. Like Ford's assembly line, to which it bears a close relationship, the supermarket arose not by design but rather as the not-wholly-expected result of the confluence of many different processes that had evolved simultaneously. Some of these forces had little direct connection to the packaging and marketing of food.

The automobile was itself a crucial factor, because it made possible large stores serving geographic areas far larger than those served by stores that depended on a pedestrian trade. People buy more when they don't have to worry about how they are going to carry it home. During the 1920s, American automobile ownership rose from nine million to twenty-six million, and the number of miles of paved roads tripled. Neighborhood stores offered people what they needed; supermarkets offered a great range of things that people didn't know they wanted until they saw them. The old-fashioned convenience and familiarity of the corner store gave way to a wide choice of products at low prices with the extra convenience of a parking lot. Suburbia was a very popular option for those who could afford it during the 1920s. And it is no accident that the large combination grocery-meat-produce stores that were the precursors to the supermarkets emerged first in California and Texas, many of whose cities were taking on the loose-knit form the automobile demanded. During the 1930s, suburbia became a goal for reformers, a way to empty out the overcrowded, substandard housing of the big cities and provide light and air for the masses. This goal was achieved, more effectively than many of its erstwhile advocates wished, during the mass suburbanization of the post–World War II era.

Home refrigerators also played an important role. They had been introduced as a luxury good during the 1920s, but the greatest expansion of the industry came during the 1930s. The single biggest factor was the introduction of the Sears Coldspot, at a price significantly below its competitors, and with Loewy's smartly styled body. Although the Depression was clearly not a period of great material expansion, the era was suffused with propaganda to encourage people to start consuming again. The refrigerator proved to be an indulgence that many could afford, and it became standard in American homes during the 1930s.

The refrigerator freed shoppers from the necessity of daily food purchasing, while the automobile allowed shoppers to drive past the corner store to markets with larger selection and lower prices. Both technologies permitted people to separate themselves from their immediate surroundings and replace human relationships with direct, impersonal contacts with packaged products.

There is little doubt, however, that packaging was what made the supermarket not only possible but close to inevitable.

For most neighborhood grocers, the primary value of packaging was not promotional, because their stores generally had little room for display. As we have seen, most retailers were, in fact, reluctant converts to selling in packages because it represented a loss of control over their purchasing and pricing. And when they did begin to stock packaged goods, the advantage for retailers was that packages made it easier to manage the store. They reduced spillage and waste, made cleaning easier, and cut down on the amount of time spent weighing and wrapping. Because packages stacked compactly, grocers were able to offer a larger stock of goods than they had before in the same space.

The first store that sold only goods in packages had opened in New York in 1907. Packages were also essential to the A&P economy-store system, launched in 1913. These stores, which in 1924 averaged six hundred dollars a week in revenues and fourteen dollars a week in profits, were such lean operations that, for many, even the incompetent grocery boy from the Uneeda advertisement would have been an indulgence. Their operation depended on one clerk in a room full of packages. In 1924 the typical A&P stocked six hundred items.

There is little doubt that the economy stores were a tremendous commercial success. Easily replicated in rented quarters, they enabled A&P to become the dominant grocery chain with minimal investment in real estate and low labor costs. But even at that time, trends in grocery manufacturing and retailing suggested that this minimalist approach was only a transitional phase. The personal touch of the grocer-proprietor had been removed, but products were given few opportunities to assert their own personalities. Moreover, the Aunt Jemima Effect was leading to a proliferation of new, desirable, profitable products, more than the little economy stores could possibly manage.

The challenge was to develop a system of retailing that was as rational, manageable, impersonal, and friction free as the systems by which the goods were produced. In 1916 the Memphis grocer Clarence Saunders developed the heart of such a system in his first Piggly Wiggly store. The system, which he patented, reconfigured the grocery store as a maze. Once shoppers passed through the turnstile, they were directed down one aisle, up the next, and on through the third and fourth until they had passed every item in the store, before arriving at the cash register and exit. This system industrialized consumption by making shoppers assemblers of their own order. Like bottles being filled or Model Ts being assembled, shoppers who went through the turnstile entered an efficient, inexorable process.

Some early advertising for the Piggly Wiggly self-service concept made a virtue of its impersonality by playing on the shopper's lack of confidence. The store freed the shopper from having to confront the grocer with a series of clear choices. Rather, she could wander about the store, freely, and let the products on the shelves help her develop her own ideas. The packaged products would help her clarify her own desires, and she needn't worry about the grocer's disapproval or the impatience of those behind her at the grocer's counter. This system created an environment in which packages could speak for themselves. People were exposed to packages that reminded them of things they had forgotten they needed. More important, the environment exposed people to enticing packages of things that people didn't know they wanted. And while a self-service store would require more floor space than a grocery consisting of a counter and a stockroom, it could translate into more sales and fewer labor costs per square foot.

Impulse buying had probably always been part of human behavior, but such a system made it possible to understand it, quantify it, and profit from it. It obviously owed quite a lot to the idea of the cafeteria restaurant, introduced in 1895. Nevertheless, the idea that people might be attracted to a can of soup in much the same way as they might be to soup they could see and smell on a cafeteria line was not obvious. But this insight proved to be powerful. By 1922 there were more than 1,200 Piggly Wigglys, either owned or franchised by Saunders, organized on this principle.

Saunders gave his customers wire or wicker baskets in which to place their purchases. Thus gravity emerged as a key limit on desire. Although shoppers might have wanted more than they could carry, their muscles provided a strong signal that they were buying too much. In 1919 a Houston store, Henke and Pillot, developed an industrialized variation on the Piggly Wiggly scheme. The store's shelves were configured in a giant M and were stocked from behind. Shoppers had baskets that rolled along tracks built into the floor.

This system brought customers closer to every product than Saunders's plan had, but it proved too demanding and inflexible for widespread adoption.

In fact, even without such tight systems as Saunders's, the combination of packages and automobiles was encouraging groceries to grow and add more self-service counters. Much the same thing was happening to chain drugstores, hardware stores, and retail establishments of all sorts. There were shoppers who declared that assembling their own groceries was demeaning, and grocers who feared onslaughts of "tomato squeezers." Still, as Woolworth had already demonstrated, giving customers direct access to the goods, letting them pick up the package and read the label, moves more products out the door. It's difficult to define just when a large, self-service store turns into a supermarket. But with their large stocks and even larger parking lots, supermarkets were appearing throughout the West and Southwest during the 1920s.

Like industrial design, supermarkets arose in the 1920s atmosphere of abundance and business boosterism but took hold during the Great Depression. Two events in the New York area in the early 1930s really brought supermarkets to public attention.

"Pile it high. Sell it low." This was the slogan and philosophy of Michael Cullen, who in 1930 opened his pioneering supermarket in a sprawling former garage in Jamaica, Queens. King Kullen was a low-overhead, price-cutting operation well suited to Depression-wracked consumers. Although, at about six thousand square feet, it would be considered a modest-sized supermarket today, it was five times the size of the large, new combination grocery-produce-meat stores that were then spreading through the Northeast. There was virtually no service. Some customers got lost, which was all right from Cullen's point of view because in their searches they found things that they weren't expecting.

The customers were without clerks but among friends, the nationally branded packaged products that Cullen was piling so high. "Pick your own national brands," said the advertisements. "Serve yourself and save." Customers felt no relationship with Michael Cullen, merchandising master though he was, nor were they likely to know the checkers or be expected to explain their purchases to anyone in the store. He allowed the manufacturers' packaging and advertising to do their selling jobs, unmediated by anyone. He was even able to benefit from customers' hostility to chain stores. Even though there were fifteen King Kullens by the time Cullen died in 1936, it was understood, especially at the beginning, as something completely different from the A&Ps and Grand Unions that people knew and, for various reasons, resented.

Like King Kong, who came to New York shortly after, King Kullen represented a change of scale that transformed the very nature of the beast. That was even more true of a store called Big Bear, opened in a vacant automobile plant in Elizabeth, New Jersey, in 1932. It was, in many ways, a precursor to today's warehouse shopping clubs. In an enormous interior, packaged grocery items were sold directly from their shipping cartons at very low markups. Large, aggressive advertisements sounded a populist note, promising "no high-salaried executives," "no large overhead," "no fancy frills and fixtures." Other parts of the plant were sublet to concessionaires who offered household goods, produce, and other items. The rental of these concessions was, in fact, the generator of profits. Groceries were sold at or below cost, in an effort to build traffic and create value for the concessionaires.

Unlike King Kullen, which was conceived as a chain, Big Bear was one of a kind, but it was an unprecedented kind. Within a year after its opening, the store's sales were totaling a hundred thousand dollars a week, which was equivalent to the total sales of all 100 A&Ps that were then operating in the Newark-Elizabeth area. Not surprisingly, numbers like this got the attention of A&P, which began experimenting with self-service bargain basements and with a price-cutting program the company called Baby Bear. In 1936 the chain launched an experimental program to open 100 supermarkets. By 1938 the company was operating more than 1,100 supermarkets, typically after closing 6 to 8 stores in the market area first. Other big chains followed, though not as rapidly.

The final key ingredient in the creation of the supermarket came in 1937, when Sylvan Goldman, an Oklahoma City grocery chain owner, invented the shopping cart. Goldman's insight was not so much the idea of putting groceries on wheels, which had been suggested before. The difficult part was to reduce the amount of space the carts took up when they weren't in use. Goldman's first model was essentially a folding chair on wheels to which a second seat had been added. Shopping baskets fit above and below. When he introduced them, the carts were not immediately successful. Women told him they were tired of pushing babies, while men thought that pushing a cart would make them look weak. Only after he hired shills to push the carts through the store were customers willing, and they proved very willing indeed. Goldman started a company to make the carts, and his subsequent improvements included the design of a child's seat, and the development of the now-universal design of a single large basket with a pivoting back wall that permits them to nest. A diabolical variation, introduced in 1952, was a child-sized cart into which children who were playing shopper placed items that were often kept and paid for at checkout time.

The allure of one-stop self-service shopping was obvious for the consumer, but its benefits for manufacturers were perhaps even greater. As Alvin E. Dodd, president, American Management Association, wrote in 1936: "Manufacturers realized that a 'mighty little man' of glass, or wood, or paper, or metal was ready to become their most vigilant salesman, a salesman who never fell down on the job — never took time out. He was the package — with a bolt of sales lightning in each hand and with a general air of 'buy' emanating from him."

Self-service gave manufacturers unprecedented control over how their products were sold. Because the grocer no longer came between the product and the buyer, companies were free to integrate advertising and packaging to give each product a distinctive personality. "By eliminating store clerks, self-service gave consumers an uninterrupted opportunity to make their own choice," said Paul Willis, longtime director of the Grocery Manufacturers Institute, in an interview published in 1986. "By displaying the manufactured products instead of hiding them behind the counter, supermarkets gave branded products the opportunity to be chosen."

"The supermarket created a more objective position for everyone — the manufacturer, the business and the consumer," said Joseph P. McFarland, a longtime executive of General Mills. The objectivity to which he referred was the elimination of the clerk and his suggestions or perceived disapproval. Removing the human element from selling is perhaps the key insight of self-service and, indeed, the logical consequence of packaging. Every transaction between people has a complex emotional component. Both parties are extremely sensitive to the words, motions, and eye contact of the other. In essence, they decide whether they like, respect, trust, or even desire the other in the course of the purchase of a loaf of bread or a box of cereal. This activity is usually mildly enjoyable. Nevertheless, personal involvement creates friction in the retailing process not because it is unpleasant, but because it slows down the sale. Even a clerk that the customer likes and respects can get in the way of a purchase if the buyer isn't sure whether the choice is a wise one. Mostly, though, the problem is that all this personal interaction is demanding. It leaves little time for uncertainty. It requires a kind of emotional intensity that most people would rather not lavish on a trivial purchase.

In touting the objectivity of the supermarket, McFarland was clearly not referring to rational thinking on the part of the shopper. According to a 1938 survey done for Du Pont, about 21 percent of the items bought in stores with sales clerks were purchased on impulse, while nearly 33 percent of those bought at self-service stores were entirely unplanned. Even more significant were the total grocery bills for each type of store. At an independent grocer, the total purchase averaged thirty cents, for chain stores sixty-five cents, for small self-service stores a dollar, and for the largest supermarkets three dollars and fifty cents. Much of this disparity must have derived from the frequency of visits to the respective types of store. The nearby grocery was handy for an item or two, while a trip to the Big Bear was an expedition. But it seems clear that the proliferation of self-service stores was increasing the volume of goods sold.

Supermarkets were designed and lighted to let the packages shine. Groceries had been small and cramped, and even the Piggly

Wiggly stores that pioneered self-service depended on constricting the customer. In contrast, the pioneering supermarkets appeared high and airy. They provided a feeling of freedom and, above all, abundance.

One important visual feature of 1930s and 1940s supermarkets and large groceries was the "package pyramid." This was a collapsible construction of dummy packages, resembling a skyscraper of boxes, that asserted the brand's presence in the store. Real boxes of the product were clustered around the pyramid's base. The temptation was to grab one. The package pyramid was a characteristic product of the visual culture of the 1930s. Like the megalomaniacal dance numbers from Busby Berkeley musicals, or for that matter the mass rallies of Hitler and Mussolini, the supermarket built countless small elements into a composition that was at once variegated and abstract, almost unimaginably large, emotionally overwhelming, and mostly empty.

Supermarkets were, of course, part of a long-term cycle in which price-cutting retailers arose to challenge dominant forces in the industry, many of which, like A&P, had originally prospered as price-cutters themselves. Although A&P, Kroger, Safeway, and the other important regional grocery chains eventually converted to supermarket chains, the first supermarkets were independents that threatened the chains' power. During the 1910s and 1920s, many manufacturers of brand-name products resisted the chains' price-cutting practices and sided with the independent merchants who gave them credibility. This time, the grocery manufacturers were on the side of the price-cutting supermarkets. One reason was that the chain stores were very important competitors. A&P's private-label brands were among the most successful in the country, and Arens's packaging for them was sophisticated and more effective than that for many national brands. Still, the key change that had come in those two decades was that the national brands no longer needed the endorsement of the neighborhood grocer. Instead, the new supermarkets won their credibility by featuring and promoting the nationally advertised packaged brands.

During the 1920s, and especially during the 1930s, lobbyists for independent grocers fought a campaign in Congress and in the state legislatures to curb the growth of chain stores, and twenty-seven states passed laws imposing high taxes or other restrictions on chain-store operations. The stores were depicted as enormous monsters from out of town who put your neighbors out of business without making any contribution to the community. Many people did find the new impersonality of buying to be upsetting, and the presence of giant companies on Main Street did signal a shift toward less self-sufficiency and an increasingly centralized economy. There is some evidence that people voted one way and shopped another, both supporting the anti-chain-store movement and shopping in these stores for bargains.

The taxes and regulations proved short-lived. In some cases courts struck them down, and in others state legislators learned that voters didn't want to be protected from the specter of lower prices and repealed the tax. Most packaged-product manufacturers stayed out of the public debate, but in general they sided with the independent grocers who kept their prices high and predictable, rather than with the large chains who were capable of pushing their profit margins down. A&P was the target of a landmark antitrust case that dragged on for many years to a determination of quite limited guilt. Government seemed just as ambivalent as many shoppers. Even as the Justice Department moved against A&P, social workers ordered welfare recipients to shop at chains where they would get more food for their money.

Packaged products were unquestionably a major force in the move toward bigness and the loss of human contact in retailing. This really could not have happened without them. But although resentment against chain stores smoldered for most of two decades, albeit without many long-term consequences, almost none of this animosity was directed at the packaged brands. Indeed, because they supported legislation to fix retail prices, they were viewed as allies of the independent merchant. It seems likely, though, that most people did not see packaged products as part of the problem. They were not perceived as part of destabilizing changes in the society but rather as entities that were firmly anchored, reliable, and friendly.

The only challenge to packaged products came in 1933, when the so-called Tugwell Bill proposed a uniform quality-grading system to be displayed on all labels. "Grade-labelling destroys trademarks by making brand names meaningless," the Grocery Manufacturers Association argued. This would result, the manufacturers argued, in the loss of benefit from three decades of advertising, packaging, and promotion. They also argued that grading "cuts the line between the consumer and the manufacturer as the identified source for the product." They did not seek to justify the sanctity of this link, which patent-medicine makers and others had frequently abused, nor did they weigh it against the importance of the link between the citizen and the government as protector of the public welfare. In any event, the Tugwell Bill was amended to provide for minimum standards, but not for grading, and it passed.

If such grade standardization had passed, it seems unlikely that it could have made very much of a difference. It assumed a world of commodities, one of which is replaceable with another, while the logic of packaging was moving toward products made from multiple elements that are not directly comparable. Distinguishing a grade-B salad dressing from a grade-C salad dressing would have required a heroic act of bureaucratic imagination. There is little doubt it would have been done, but manufacturers still had at their disposal the tools of packaging and advertising that could play a far greater role in how the consumer perceived the product.

We can see an indication of how the manufacturers would have coped in Arens's design for A&P's bottom-of-the-line Iona canned fruits, introduced not long after the Tugwell Bill debate. The colorful fruits were set off vividly against a black label, an almost unheard-of color for a food product at the time. There was an implication of luxury in the use of black, which was used in a similar way in upscale products of the 1980s and the successful President's Choice and Master's Choice private labels of the 1990s. In a prominent place on the Iona label was the description "Grade C," which challenged, but did not negate, the overall message of the label. The shopper could tell from the price that she was buying the cheapest

138

available can of pears, and during the Depression, she might have faced a choice between these pears or none at all. What the package's design did was reassure the buyer that even this lowest-price choice was a respectable one that would not betray the trust of those to whom it was served.

6 *A world of packages*

American soldiers went to fight World War II wearing can openers around their necks.

As Napoleon had foreseen, and Queen Victoria's navy had quickly grasped, preserved food made it possible to lengthen supply lines and fight in many widely scattered, distant places. The unprecedented geographic range and scale of this war, fought on three continents and across the world's oceans, was made possible by a number of technological advances, particularly in radio communication and aviation. Canning was a relatively old technology, but the can opener, which rattled against a soldier's chest along with his identifying dog tag, indicated that food packaging was an essential part of fighting the war.

The GI can opener — a hinged piece of metal about an inch and a half long and an inch wide — is not a recognized classic of modern design, though it probably ought to be. It was a truly minimal tool, the least you could possibly use to get the job done. Yet, especially when its sharp-pointed nose was folded outward for use, the device had character, a friendly-serious aspect appropriate for something carried so intimately and so necessary for survival.

World War II was the last big can-opener war. Since then,

140

soldiers' rations have moved into plastic containers, plastic and foil pouches, and other novel sorts of packages, many of them first used on a limited basis during World War II. The war represents a dividing line between a culture of tin cans and a culture of plastic squeeze bottles. (In fact, the first mass-marketed squeeze bottle, Stopette deodorant, was introduced in 1945, shortly before Hiroshima.)

And in part because postwar planners wanted to keep America from being vulnerable to the kind of aerial bombing that had devastated cities in Europe and Japan, the postwar landscape was transformed. Cities and towns, with their Main Streets and corner stores, gave way to a sprawling new suburban landscape of separated houses and shopping centers with huge parking lots. Wartime propaganda promised soldiers and workers on the home front that their sacrifice would be rewarded with a new way of life. These promises were kept. Americans have lived in and elaborated on this postwar vision for nearly fifty years.

Many of the trends we have already examined accelerated in this postwar world. The supermarket quickly changed from a novelty to part of everyday life. Transactions that involved other people became rarer. Packages developed more personality, and they were adapted to please all sorts of people. New kinds of packages, such as aerosol cans, spawned whole new categories of products. And the fundamental promises of packaging — standardization and protection from danger — were adapted by businesses that ranged from fast-food restaurants and tourist destinations to dental clinics and income tax preparers.

If we look at America's last half century from the point of view of what was offered for sale and what people bought, the trend has been steadily toward more. Each year, the number of packaged products has grown larger, as have the stores in which they are sold. Even the individual packages have grown larger. Several recessions, a twenty-year stagnation of personal income, two energy crises, two outbreaks of environmental consciousness, and a massive influx of women into the workforce have not significantly altered this fact. (The increasing number of working women appears to have helped sustain this material expansion of American life and enhanced the market for products that promise to be timesaving and convenient.)

141

Thus, it makes sense to look at the war and the years since as a single period, despite its vastness. The rest of this chapter will trace some of the consequences of the war and its immediate aftermath, while the chapters that follow will abandon the historical approach and consider such issues as design, psychology, consumer protection, and the environment.

You can still open a can with the World War II GI can opener. Today's cans are thinner than those used during the war, and those were thinner than cans had ever been before.

The late 1930s had brought a re-engineering of cans to reduce the amount of tin used in them. North America lacks tin, and most that was imported came from Malaya, which had passed from British to Japanese control. When the United States became involved in the war, there was good reason to reduce the amount of steel in cans so that the metal could be used in ships and tanks. Thus, cans were forced to become more efficient in their use of materials, while remaining sturdy enough to be transported. This thinning of cans continued a process that began during the Civil War eight decades earlier and continues today as package manufacturers and grocery marketers face environmental challenges.

Such increased efficiencies are an almost inevitable result of war. Materials become more precious in wartime than in peace, and there is great incentive to improvise. The fifty-five-gallon oil drum, the container that carried the stuff that made the war go, had countless secondary uses. It was durable and plentiful, adaptable to make everything from emergency bridges to solar-heated showers. A whole culture of ingenious reuse grew up around it, which unquestionably had an impact on the postwar mania for do-it-yourself household products. (The GIs were not the only ones inspired by the oil drum. In the Caribbean it gave rise to steel-band music.)

Wars also tend to speed the realization of inventions that had existed only as concepts before. The Manhattan Project to develop an atomic bomb, the advances made in computer technology in connection with code breaking, and the development of radar are

142

the best known of these from World War II. But there were other less secret, less decisive breakthroughs that have nevertheless had an important impact on contemporary life. Many of them can be summed up in a single word: plastics.

That's a word that describes a great range of materials with different chemical compositions and different properties. Although polyethylene, which in all its forms has been the world's most used plastic material, was introduced on the eve of World War II, plastics were not new materials. They had been around since the turn of the century, and they became part of ordinary life after Kodak introduced its first Bakelite camera in 1911. The production of plastics by weight had been tripling every five years since then.

The first plastics had been produced as substitutes for natural products, such as ivory and tortoiseshell. Later, their fluid, moldable properties gave rise to an aesthetic of its own that was consonant with the streamlined, Moderne style popular during the 1930s. Plastics were considered up-to-date, and they had an aura of affordable glamour. A famous semicircular promotional box used to introduce Kool menthol cigarettes was stylistically akin to the proscenium of Radio City Music Hall and was similar, as an object, to the streamlined Catalin radios of the same period. But it did nothing for the cigarettes, except make them more attractive. Similarly, as we have seen, cellophane was sold on the basis of its sparkle, its transparency, and its implication that the product within was pristine, rather than for its ability to keep moisture out or for its light weight.

By 1940 the plastic-packaging business grossed about $500 million, which represented a tiny but growing portion of the entire packaging industry. Moreover, the companies that were interested in the continued expansion of consumer use of plastic, such as Dow Chemical, Union Carbide, and Du Pont, would inevitably play a major role in any war effort. Nevertheless, because plastics were used primarily for decorative caps, special promotional packs, and windowed folding boxes, it is understandable that bureaucrats charged with converting the country to a wartime footing would find them to be expendable. In November 1941 the federal government issued

143

a materials-limitation order whose practical effect was to put the plastic-packaging industry out of business for the duration of the anticipated war.

But there was a counterargument. Shortages of metals during World War I had, in fact, forced manufacturers to consider using plastics for caps and other types of closures. This was the beginning of the plastic-packaging industry. Perhaps plastics could be used in this new war for military purposes, to ease delivery of supplies and to replace materials for which there were other, more urgent uses.

At the time of the 1941 order, it was already evident that tin was in short supply and that other metals would become scarce. The United States had abundant oil reserves, and the polymers for manufacture of plastic resins could be produced along with other petroleum products. Plastics manufacturers exhorted the government to reconsider its order. In return they were challenged to rethink the nature of their materials because, for war purposes, the allure and display value of plastics was irrelevant. Plastics had to be justified on purely functional grounds. Surprising as it seems, this was something plastics manufacturers had not done before. Moreover, the conditions of war and military standards that were written with metal components in mind required that plastic caps show far greater shock resistance than was necessary for nail polish or toothpaste. It was essential that rigid plastic boxes containing small spare parts not shatter in the annoying way that similar boxes, used for face powder, often had. Typically, military items went through eighteen separate handlings — from manufacturer to supply depot, railcar, coastal depot, ship, and advance base — before reaching a destination in the combat zone. Items had to survive air drops into the ocean and periods of storage in places that were not protected from the weather.

The influential companies that dominated the plastics industry convened a meeting among themselves to consider how plastics could be used in warfare, both to replace steel, tin, wood, and rubber and to perform tasks for which plastic materials were better suited than any other. They were successful in overturning the order that would have shut their industry down, and they were also able to work with the military to create hundreds of uses for plastic packaging materials.

One of the most obvious and best known was an ethyl-cellulose army canteen with a screw cap, whose light weight and rounded shape made it easier to carry than metal canteens. Some firearms were shipped in air-filled plastic bags, so that they would float when dropped in water. Every GI was issued a clear plastic envelope in which to hold his paybook, and similar plastic covers were used to protect maps, blueprints, repair checklists, and other essential paper documents that had to be referred to frequently. A process was developed to dip metal machine parts in plastic, which kept the steel free of corrosion and could be peeled off only when the parts were needed.

Rigid plastics were used for the shipment of drugs, ammunition, and spare parts and for the housing of individual first-aid kits issued by the navy. Plastic-coated fabrics were used for making clothing and blanket bags. Plastic films, laminated with paper and metal foils, were used for wrapping coffee and many other food items. It was during World War II that Americans came to expect a cellophane wrapper on their cigarette packs.

Many of these military uses for plastic materials were adaptations of earlier civilian uses, while others gave rise to later nonmilitary uses. The specific items produced are, however, less important than the change of perception of plastics, first by the companies that produced them and then by others.

In the half century since World War II, plastics have never recovered their Depression-era chic. But the increase of plastic packaging continued at a brisk, steady pace, not because consumers were demanding it but because it was very functional for producers, transporters, and retailers. Plastic containers can be colorful and emotionally appealing. But such financial and operational virtues as light weight and shatter resistance are the real reasons plastic packages are everywhere (including a lot of places that they shouldn't be).

The proliferation of plastic packaging after World War II was not as explosive as was that of paperboard folding boxes seven decades earlier, and some of the most important plastic packages —

such as the two-liter soda bottle — came more than twenty-five years after the war. The growth of plastic packaging has, however, been steady and long-term. The perception that plastics are more artificial and somehow less trustworthy than such older materials as glass, tinplate, or cardboard has probably slowed consumer acceptance of plastics. More recently, plastics' resistance to natural forces of decay, which was once an asset, has come to be seen as a problem. Nevertheless, plastic packaging is steadily on the increase, and even packages made predominantly of other materials, such as glass jars, often have a plastic collar around their caps to reassure shoppers they have not been tampered with.

It is difficult even to talk about plastics, because they constitute such a wide range of materials with an even wider range of properties. Even a particular plastic resin can offer different sorts of strength, moisture protection, flexibility, and appearance, depending on how it is processed. Plastics are also present in such packages as the gable-topped "cardboard" milk carton, and they are frequently laminated with paper, metal foils, and even other varieties of plastic. While most people have a vague idea that plastics are very complicated strings of molecules, an emotional grasp of "what they really are" is elusive. Unlike wood or paper or even steel, plastics don't have character, but, rather, multiple personalities. The right question to ask about a plastic is not what it is, but what you can make it do.

The World War II experience set the stage for an approach that can custom-tailor the package to suit its contents. With such older packaging methods as canning, it was necessary to alter the contents of the package to suit the technology. In the case of canning, this meant that nothing that couldn't be boiled could be put in a can. Now it is possible to specify an internal environment, a degree of strength and rigidity to withstand transport and warehousing, and an outer surface for printing, then to engineer a package to do the job. It might not be made entirely of plastic, but it is likely that there will be at least one plastic layer playing an important functional role. In reality, cost and production requirements force

A

This nineteenth-century soap (A) uses four different typefaces for the name of the product. After 1900, packagers simplified and focused their packages. Wrigley's arrow (B) is a symbol from that era that's still potent.

B

With its seals, endorsements, garlands, and flowery prose in fine print, the Budweiser can shows its roots as a Victorian label.

The center seal represents the continents. Anheuser-Busch was the first brewer to use the railroad system and emerging advertising media to market nationally. The brand identity easily survived Prohibition.

Like nearly all beers and soft drinks nowadays, Budweiser comes in an aluminum can, whose body is a single unit, the top another. The body tapers a bit at the top to save weight and cost.

Beer doesn't require a nutrition label, but the health warnings are printed in the condensed, lightface type packagers use to discourage reading.

Recycling is economical for aluminum cans, and recyclers like them because they bring a high price.

The can is the same front and back, which is common in packaging for beer, cigarettes, gum, candy, and other habit-driven products.

Gerhard Mennen launched the toiletries business with an innovative tinplate can (A), with a baby on the front to attract mothers, and himself on the cap to reassure them. Lyon pioneered tooth powder, then diversified into a tube (B), while Colgate, the tube pioneer, dispensed toothpaste in ribbons (C).

Aerosols for toothpaste (D) didn't catch on. Current packages include stand-up plastic tube (E), Mentadent's lavish double pump (F), and Tom's of Maine's old-fashioned box and tube (G) with its chemical analyses, "cruelty free" assurance, and discussion of recycled content.

A

B

C

D

E

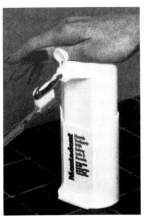

F

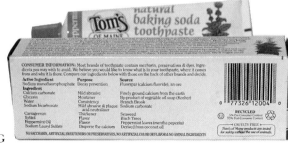

G

A

B

C

Mid–nineteenth-century cans were largely handmade and often had simple labels (A), but they soon sported colorful pictures of their contents (B,C). Campbell's simplified label (D), introduced at the turn of the century, emphasized the brand, not the contents. Campbell abandoned the austere red and white, first for its international lines (E) and more recently domestically, to compete with the appetizing pictures on rival products (F). Simple labels now express bare-bones economy (G).

D

E

F

G

A

Marlboro, first marketed as a women's brand (A), had a sex-change operation in the 1950s (B), of which the distinctive red-and-white, hard-sided, flip-top box was an important aspect.

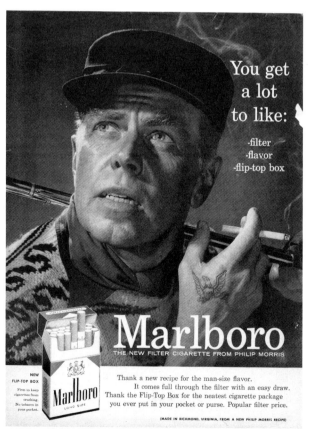

B

15

A

B

C D

E

Patent medicines were the first widely recognized and distributed packaged product. Most were known by the shape of their bottles rather than the contents (A), and graphic devices, a safe, for example (B), were used to communicate safety. Poisonous substances were conventionally placed in bottles with sharp points (C).

Swaim's Panacea (D) was a dangerous concoction with ingredients that included wintergreen and arsenic. Swaim's commissioned a book that illustrated a miracle cure (E).

A

B

D

C

E

The contour, hobble-skirted Coca-Cola bottle was created in 1915 and has become the world's most recognized package (A). Tab-open tops, introduced during the mid-1960s, finally led Coke to move to cans, which didn't succeed in changing the product's imagery (B). The "dynamic ribbon" motif is credited with keeping the contour image alive even on today's standard containers (C). A crude version of the contour in PET plastic (D) is reintroducing the image in the United States. In China, a reusable glass contour bottle (E) has characters that make Coca-Cola into a pun for "easy mouth, easy happiness."

compromises, and relatively few products are sold in their absolute optimum packages. But the ability to design the package as a specific environment has increased producers' freedom in developing new products. For example, the fat-free cakes and bakery products introduced in recent years depend on an outer, clear plastic film that keeps the product's moisture inside the box. Fat is what keeps most cakes moist. The new ones depend on packaging.

World War II prompted similar technological shifts in a number of different industries and companies. These, along with the prospect of a greatly changed postwar world, induced some firms to change the way they saw themselves. Companies came to identify themselves less with their raw materials and the technologies they used to process them, and to look more at how their products were used and how they were marketed. The wealth of a company was increasingly seen to lie less in its production base than in its market base. Rather than try to sell people what you made, you could make what people would buy, building on your capital of consumer trust.

This change came to Procter & Gamble — an old company with several dominant products in the marketplace, inventor and impresario of the soap opera — in a paradoxical way. The company had, from the time of its beginnings as a candle and soap maker, been identified with fat. At the beginning of World War II, its two best-known products, Ivory soap and Crisco shortening, continued that tradition. The powerful explosive nitroglycerin is not the sort of thing you find around most kitchens, but it is a fat. The company's expertise in processing fats led the government to ask Procter & Gamble to build and manage a plant to make explosive shells and other ordnance. In the meantime, shortages forced the company to stop producing several of its products.

Procter & Gamble decided to continue to advertise these discontinued lines during the war and to identify itself less as the top company in the fat business and more as one of the most trusted purveyors of products for the home.

Synthetic detergents had been invented in Germany during

147

World War I, precisely to respond to the diversion of fat supplies to military uses. Procter & Gamble had been in the business in only a small way, with a powder called Dreft, introduced in the 1930s as a mild cleaner for dishes and delicate garments. But largely because Procter & Gamble and all its largest competitors worldwide were committed to fat-processing technology, soap had not been seriously challenged. Such a challenge was probably inevitable, and Procter & Gamble decided to keep control of the family wash by developing the products that would supplant its own.

The first major result of the effort was a heavy-duty laundry detergent, Tide, introduced in 1947. In countless "new and improved" manifestations, most recently liquid and superconcentrated powder, Tide has been the leading American laundry detergent brand for close to half a century.

The orange-and-blue, targetlike Tide box, designed by Donald Deskey, who before the war had been known for his fine furniture and the interior of Radio City Music Hall, is an indisputable landmark in package design. It is hardly an exaggeration to say that the box itself was formulated as carefully and rigorously as what was inside.

Its bull's-eye motif recalls that used for Procter & Gamble's soap-based laundry powder Oxydol. Thus, it exploited a motif in which shoppers already had a high measure of confidence and subtly identified the product's category. The risk was that the new product would cannibalize sales of the old, something that was probably inevitable no matter what the package. Besides, the differences between the Tide and Oxydol packages are more significant than their similarities.

On the earlier box, the bull's-eye centered on the *O* in Oxydol, and the blue and orange circles are clearly meant to evoke a target. On the Tide box, the concentric circles are yellow and orange, related, high-intensity colors that seem to undulate. They are less likely to be perceived as a target and more as waves or lines of force. Both of these interpretations reinforce the name of the product, which speaks of something both wet and inexorable.

On both boxes, the outermost circle is not contained within the boundaries of the box, a graphic device that induces viewers to

148

complete the circle in their own minds. This tendency, though sub-conscious, nevertheless causes viewers to participate in the dyna-mism of the package design. Moreover, when viewers extend the design in their own minds, they are also expanding the psycholog-ical impact of the package.

The Tide design involved some sophisticated color research. The initial challenge was to communicate that the new product would have sufficient power to handle heavy laundry. Procter & Gamble's Dreft and other detergent powders on the market had been sold on the basis of their mildness, and Dreft's package, fea-turing thin, Moderne lowercase lettering, had communicated this message extremely effectively. Orange had been shown to be a pow-erful, heavy-duty color, the color of warning signals. It communi-cates precisely the opposite of delicacy. Yellow is the most visible color, one that simply cannot be ignored. But while these two strong colors may be saying "Keep out" to dirt, could this novel laundry product also prove hazardous to your clothes? That is the purpose of the blue lettering in which the word "Tide" is written. Blue is a memorable, likable color, which symbolizes softness and mildness. The box seems at once to be promising power, with its aggressive orange and yellow, and restraint, in the blue in which the name of the product is written. The sensitive, authoritative blue stands in front of the powerful but unrestrained orange and yellow to com-municate the message that the product is as strong as it needs to be, but not a brute.

Two years later, when Procter & Gamble brought out an-other heavy-duty detergent, Cheer, the effectiveness of detergents had been established, and the package was made predominantly blue, to communicate mildness, punctuated with streaks of bright red and yellow, which suggested vivid-colored clothes hanging on the line and communicated the power of the contents. Thus, the shopper can choose between Tide, whose package was designed to signal that it was powerful yet mild, and Cheer, whose image is mild, yet powerful. A few years later, Procter & Gamble reformulated Cheer with whiteners that made the powder blue. To compensate, the blue on the outside of the box was reduced and more white was added.

149

The successful introduction of Cheer demonstrates how Procter & Gamble's control of some of the most powerful products on the market gave it the muscle to get more shelf space. During this period of expanding stores and expanding incomes, Procter & Gamble launched a number of products, like Cheer, which sought to win the number-two market position away from competitors' brands. (During the early 1990s, the increasing clout of retailers and poor performance of some of the secondary brands forced the company to reverse this practice.)

One of the most interesting and radical of Procter & Gamble's variations on an existing product was its second entry in the shortening category, Fluffo. Crisco, the long-dominant brand, had undergone a redesign to make the blue-and-white can look crisp and clean, part of the modern kitchen. The whiteness was associated with purity, the blue with a scientific form of authority. With Fluffo, the shortening was itself "golden," and the can was covered in a checked-tablecloth design that was homespun and warm. While Crisco tried to look like something a dietitian might recommend, Fluffo's implied endorser was a grandmother. Did you want to see fat as part of a balanced diet, or as an expression of love? The consumer could choose, though the cans were more different than what was inside.

The Procter & Gamble move to detergent technology had much greater impact than the creation of a long-running megabrand. It also set the stage for a tremendous number of new products, including shampoos, liquid detergents, dishwasher detergents, and household cleaners. Most large manufacturers had offered several varieties of soap on the grocer's shelf. But detergents can be formulated for countless tasks and be marketed to solve very specific problems for well-defined groups of people. The insight that packaging usefulness is more profitable than packaging a specific substance goes back at least to Aunt Jemima. But the chemistry of detergents and the wide-open spaces of the supermarket made this logic even more compelling.

In addition, the proliferation of specialized appliances, combined with a widespread ability to purchase them, made an enormous difference. It is no accident that the first fully automatic washing machine was introduced in 1947, the same year as Tide. Indeed, small boxes of Tide were packed in most new automatic washers, just as boxes of Cascade were later packed in new dishwashers. Changes in fabrics, floor coverings, pots and pans, hairstyles, and other materials and expectations of everyday life brought forth new cleaning agents to cope with them.

Procter & Gamble calculated that at the turn of the century a family's annual soap purchases cost the equivalent of about two weeks' average wages, while in 1956 the cost was only one or two days' worth. During the postwar years, Procter & Gamble diversified into specialized detergents, dishwasher detergent, a wide range of beauty soaps and deodorant soap, deodorants for men and women, shampoos, home permanents, peanut butter, paper towels, throwaway diapers, and other products to soak up at least some of Americans' additional disposable income. Whole aisles of a typical supermarket are occupied by products that have purposes that were once fulfilled by soap.

Other appliances also generated whole new supermarket aisles and packaging challenges. Quick-frozen foods, which had been invented and first introduced before most Americans even had refrigerators, let alone freezers or freezer compartments, came into their own with the postwar appliance shopping spree. Chest-style deep freezes were touted as new necessities, and they were desirable enough to be at the center of a bribery scandal during the Truman administration. But the expansion of freezer compartments in the new, larger postwar refrigerators was what really made it worthwhile for supermarkets to install freezer cases.

At first, frozen-food packaging was really quite simple — white paperboard folding boxes, covered with an illustrative label. The freezing did most of the job of protection, so there was no need for the packages to be highly engineered, and the pictures on the boxes tended to lack the elaboration and context found on many cans of the same product. Often there were no pictures at all. Over the years, improved paper and printing processes have changed the outer appearance of frozen foods, as has the understanding that the iciness of frozen foods requires packaging that is more vivid, visually "hotter," than a nonfrozen competitor.

When frozen-food manufacturers began to understand that they could offer something that was far more than a substitute for a can, packaging began to play a far more important role. The key development here was the TV dinner, introduced in 1954. This innovation consisted of an aluminum tray divided into compartments that held a portion of meat, a vegetable, potato, and later a square of dessert. The convenience of this heat-and-eat package gave each of its components greater value than it would have had on its own. And the presumed modernity of this style of eating may also have made the food seem to taste better than it really did. In any case, it was a whole new way of packaging a combination of ingredients to establish a new, desirable product. Three- and four-compartment frozen meals are still with us, now as likely to contain a kid's breakfast as an adult dinner, and these have given rise to a wide array of frozen, microwavable entrées in plastic tubs.

Still, although frozen dinners and entrées are well established in the marketplace, this sort of packaging seems caught in a vicious circle. Consumers refuse to pay as much for food in this form as they would pay for it fresh, largely because such products are viewed as an inferior substitute for real food. Manufacturers reply that consumers are unwilling to pay enough for good frozen meals for them to make a profit, so they must devote most of their efforts to the cheaper products that sell. Supermarkets have, during the last decade, developed the selling of freshly prepared food into a large, attractive, and very profitable business, one that seems about to replace the meat department as the store's centerpiece. Food in edible form is formidable competition for even the most attractively packaged block of ice. Frozen dinners and entrées may be in danger of losing the chief advantage that packaging brings to such an assembled product — that its convenience makes it worth far more than the sum of its parts.

"All history shows no more portentous economic phenomenon than today's American market," said *Fortune* magazine in 1953.

"It is colossal, soaking up half the world's steel and oil and three quarters of its cars and appliances."

Family discretionary income rose nearly sevenfold from 1941 to 1958. A combination of younger marriages and couples who had postponed families during the Depression and war years gave rise to the postwar baby boom. Meanwhile, the extended family became less common and the nuclear family of Mom, Dad, and the kids became the norm. The number of residential family units in the United States increased 42 percent from 1929 to 1953, and families' aggregate income increased 87 percent. Between 1950 and 1956, 156 million Americans changed residence, a number approximately equal to the country's population.

Many low-paid jobs — notably laborers and domestic servants — disappeared from the economy. Those middle-class households that had been accustomed to having servants felt a loss of convenience. They sought to compensate for it by purchasing labor-saving appliances and new cleaning products, thus helping accelerate the boom. But far more significant was the newfound ability of most of those who held such jobs to move up the income ladder, mostly to well-paid manufacturing jobs. This led to a jumbling of class distinctions. Factory foremen learned to play golf; managers were not embarrassed to go bowling.

This period of increasing social and economic equality was relatively short-lived, and never so all-encompassing as it was proclaimed to be at the time. By the late 1950s, class and taste stratification was again evident in the marketplace. Still, the period of the perceived mass market made it possible to introduce great mainstream brands, such as Tide, that even today have no associations with a particular social class, region, age group, or lifestyle. They are simply part of the fabric of American life.

Suburbia, which had for decades been absorbing most of the country's population growth, was transformed. Loan guarantees from the Federal Housing Administration and the Veterans Administration, along with most other government policies, pushed nearly all residential development to the suburbs, as did the existence of cheap land and the ability to build houses in quantity. Suburbia became the most affordable place to live and the home of the most desirable consumers: those forming families and spending liberally to participate in the good life. Reaching these families became a high priority, as the amount spent on advertising increased fivefold from 1941 to 1956, and the advertising man became a glamorous figure, often played in the movies by Cary Grant.

Suburbia is the natural habitat of the supermarket, with its big parking lot and long aisles full of satisfactions and delights. In many of the new suburban areas, there were only supermarkets, no old-fashioned grocers and butcher shops, so people had little choice but to get into the habit of driving to the supermarket and stocking up.

The supermarket is, in turn, the natural habitat of the package, an ideal environment for the manufacturer to get his message across, unmediated by salesclerks. And because the supermarket tended to exist in an unfamiliar landscape, away from Mom and other traditional sources of guidance and advice, there was little to interfere with the power of a product's advertising and packaging. In the years after World War II, the producers of branded, packaged products had the upper hand over the retailers because their products were what gave their stores credibility.

The new supermarket was a land of opportunity for grocery manufacturers, who rapidly developed new products to take advantage of rapidly expanding self-service opportunities. This, in turn, sparked the growth of the packaging industry. The value of packaging materials tripled between 1939 and 1947, to $5.4 billion, reached $8.9 billion in 1955 and $14.2 billion in 1964. By 1964 Americans were consuming an average of two thousand packages a year, per person. A different calculation estimated the size of the entire packaging industry at 3.5 percent of the country's gross national product. For many products, the cost of producing the package rivaled or exceeded the cost of making the contents. In cosmetics and toiletries, for example, the cost of packaging in 1964 averaged 36 percent of the manufacturers' selling price of the product.

Design fees were an almost negligible portion of this amount, about half of 1 percent of the total value of the industry. A significant amount of package design — especially for smaller manufacturers —

154

was done by the suppliers of cans, boxes, and other kinds of containers. This had always been true to some extent, though the upgrading of such services after the war was a clear indication that package design had become more important.

"Packaging," said Deskey, "is the stepchild of advertising . . . perhaps not abused but at least neglected." Package designers say much the same thing today. But what's more significant is that there were, for the first time, design firms that specialized in packaging. Moreover, some of the industrial design heroes of the prewar era, including Raymond Loewy, Walter Dorwin Teague, and Deskey, derived a large percentage of their firms' incomes from packaging during these years. More buying power meant more products, and in turn more things to shape, package, and provide with personalities. But the designer metamorphosed from a magician to a technician, someone who understood both the chemical reactions between shampoo and container and the emotional chemistry between shampoo and shampooer.

The designer sold expertise in fields ranging from chemistry to psychology and a procedure that subjected aesthetic judgments to consumer surveys and other forms of experimentation. A client would be more likely to hire such a firm to reduce risks than to realize a vision. Deskey once told the members of a trade association that visual treatment constituted only 13 percent of the task of designing a new package. He said 37 percent of the work consisted of gathering market information; 40 percent doing technical research on materials, suppliers, production, and costs; and 10 percent convincing the client. Even the pseudoprecision of these numbers was intended to argue that the work is more engineering than art.

The post–World War II period brought marketers a glamorous new technology — television. By the mid-1950s, television had become effectively universal; just about everyone mass advertisers wanted to reach owned one. In its earliest years, when most viewers had only two or three choices of programs, television helped supply a kind of national cohesion that seemed to balance the physical separation characteristic of suburbia. Never had Americans lived farther apart and had a greater sense of connection with the center, if not necessarily with each other.

Television advertising has always been terribly expensive, which is one reason it has always gotten more attention from companies, and the public, than package design. But unlike radio, television can show the package, which makes its telegenic qualities quite important.

In the early years of television, a Columbus, Ohio, company that ran commercials featuring three long-legged dancing girls clad in oversized bags of Buckeye potato chips won praise for its imaginative use of both packaging and television. On network television, a pack of Old Gold cigarettes tap-danced weekly on the game show *Two for the Money*. Later, a king-sized pack was introduced, and the weekly dance act became a duet. Advertisers also found not very subtle ways to get their packages into the program itself. On *Double or Nothing*, another game show, the host Bert Parks and the contestants were dwarfed by a gargantuan can of Franco-American spaghetti. They looked like mice in the cupboard. On *What's My Line*, the cards that showed how much the contestant was winning for a time also featured the distinctive silhouette of the sponsor's deodorant squeeze bottle.

Television also afforded companies the opportunity to involve the package in brief minidramas. One of the most irritating series of these promoted a Procter & Gamble deodorant called Secret. The commercials featured a busybody named Katy Winters, who posed as a helpful confidant. In her hushed, deeply serious telephone conversations with her neighbors she always steered the topic to what she knew was their real problem — the fear, or perhaps the reality, that they smelled bad. "That will be our Secret," she would say, conspiratorially, once she had recommended the product that freed her neighbors from embarrassment and anxiety. The Secret roll-on package was a witty continuation of the cloak-and-dagger atmosphere that surrounded Katy and her deodorant. The dispenser was visible inside the cellophane-wrapped box, but its center was obscured by a floating white rectangle with the word "Secret" blocking a full view. With its combination of revelation and concealment of the product, this package appealed to voyeuristic impulses and hinted at Cold War anxieties about national security. The package was subtle, compared with Katy, an authoritarian who acted in the name of inoffensiveness.

The coming of television prompted a lot of redesign of existing products. Generally, this involved a new round of the kind of radical simplification that products like Campbell's soup had undergone half a century earlier. Gold medals and coats of arms disappeared, while many logos were greatly simplified. Because television was, the first dozen or so years of its popularity, predominantly black and white, those who designed packages were challenged to come up with packages whose colors and graphics would stand out sharply in the gray haze of the television screen and would still take full advantage of the power of color when encountered on the shelf.

Perhaps the most successful redesign of all time was that done for Marlboro cigarettes in 1955 by designer Frank Gianninoto. The Marlboro cigarette had existed previously in a white pack covered with weak graphic elements and a lot of copy. It was associated with women, at a time when buyers of filter cigarettes were most likely to be women. But filters were beginning to catch on with men, too, and the redesign was prompted by the desire of Philip Morris, the tobacco company, to have a filter cigarette that would appeal to all. Gianninoto's simplification was, in fact, very like the Campbell's soup can — red on top, white on the bottom, with a coat of arms that, like Campbell's gold medal, tends to disappear. The white meets the red as an arrow pointing upward, a very simple graphic device visible on even the snowiest television set.

Perhaps the most radical part of Gianninoto's design was the new physical form it gave the package. Rather than the familiar cigarette pack made of paper, with a foil liner and cellophane wrapper, Marlboro sported a cardboard box with a top that flipped open. The early advertising jingle promised "filter, flavor, flip-top box," an indication that the company thought that the box would be perceived as something special. There was a certain implication that people living a rugged life in Marlboro Country needed a tougher cigarette pack in their pockets, but the advantages of this novel pack — beyond its novelty — were never fully explained. The English design critic Reyner Banham theorized in 1962 that the real purpose of the box was to prevent people from removing their cigarettes easily from the package. "The last time a cigarette is even Brand-X is in the act of being extracted from the packet — after that it is strictly Brand Zero," Banham wrote. Opening the flip top was, he went on, "a mechanical ritual to be performed each time with the pack in view." Thus, Banham argued, the package served to remind a smoker what brand he preferred, even though "the corners of the hard box when stuffed into the traditional American shirt pocket dig into the surrounding rolls of affluent flesh every time he folds himself into the driving seat of his car."

If Banham's analysis is correct, the Marlboro hard pack was certainly an innovative way of getting the customer's attention. In any event, Marlboro succeeded, like Coca-Cola and blue jeans, as a worldwide icon of American-ness. Emerging at a moment when American cars, houses, and products were becoming increasingly elaborate, Marlboro has the stripped down, one-size-fits-all quality characteristic of the most enduring American designs.

Marlboro has succeeded so well that it has become a symbol of all brand-name products. When the brand encountered price resistance in 1993 and sales fell 5.6 percent, not only did the stock price of its manufacturer, Philip Morris, fall, but so did the stocks of virtually all other brand-name producers. Newspapers and magazines ran feature articles on the death of brands, shocked that consumers in the midst of a worldwide economic downturn should buy cheaper cigarettes as a way to save money. The *Financial World* 1993 analysis of brand values suggested that obituaries were premature. Marlboro's value had slipped 6.3 percent, leaving it at about $39.5 billion, by far the most valuable brand name in the world. (Coca-Cola was next at $33.4 billion.) A very high percentage of this value must be ascribed to the relentless advertising and promotion of the brand throughout the world. Some much smaller share must be allocated to the cigarettes themselves. But there is little doubt that the package itself — universally recognizable, never dated — deserves a good deal of the credit.

It also, of course, deserves blame for helping make smoking

a desirable activity and, especially outside of the United States, a high-status one. If Gianninoto's design was as much of a marketing triumph as it appears, it was also a public health disaster that helped induce people to smoke cigarettes who might not otherwise have done so. Everybody knows, intellectually at least, that great packages don't always hold good things. But that is a truth that the best packages try to make people forget.

The two decades that followed World War II consisted of two distinct phases. The first, which lasted until about 1953, was the period of catching up on consumption and realizing the promises of material comfort that had been made in preceding decades of Depression and war. In packaging, as in movies, cars, and interior decor, colors were clear and vivid; shapes were bold; modernity was accepted as a fact rather than as a taste. Traditional packages — glass bottles, tinplate cans, and paperboard folding boxes — still dominated store shelves. Packagers were more concerned with getting packages on the shelves than with complex refinements.

The next phase, which lasted from about 1954 into the mid-1960s, was prompted by fears that once everyone had caught up on consumption, they would stop buying and thus plunge the country back into a new depression. The tactic was to make each thing that was consumed more elaborate. In packaging, this translated into new kinds of convenience packaging, often made from new materials, that encouraged greater use of the product. The color palette for everything from cars to movies became grayer as bright primary colors were replaced by pastels. In packaging, plastics, which are well suited to these soft, impure shades, came more and more to replace harder, heavier materials.

One can get a sense of this evolution by looking at the design of Joy, a Procter & Gamble liquid dishwashing detergent for which Donald Deskey Associates did the initial 1949 design and several redesigns. The bottle was glass with a chunky, masculine profile, and the three capital letters JOY in contrasting bright colors, highlighted against contrasting squares, seemed to explode across the label. This show of strength seems to have been an attempt to overcome consumer concern that a liquid detergent would not be sufficiently powerful. As the design evolved, the glass bottle was replaced by a plastic bottle whose form was more feminine and functionally easier to handle. The letters were aligned, elongated, and the *o* and *y* became lowercase as each successive generation of the package placed greater emphasis on the product's mildness. Obviously, the glass bottle was a problem for a slippery product like detergent, and the unbreakable, graspable plastic was a far better solution. But the plastic bottle accomplished something else. In combination with the development of a narrow-holed nonclogging closure (which was a slow, difficult, but worthwhile process) the product was transformed from one that was measured out a capful at a time to one that could be squeezed onto a bit of dirt impulsively. Whenever you can shift the use of a product from a thoughtful act to a reflex, you increase the use of the product.

Something similar happened with the repackaging of soft drinks, beginning in about 1960. The original contoured Coca-Cola bottle, designed in 1915 by Alexander Samuelson, a Swedish glass-blower employed by the Root Glass Company of Terre Haute, is unquestionably one of the greatest, most beloved packages of all time. This tough little six-and-a-half-ounce bottle, which had been designed to be found by feel alone in a dark icebox or closet, had conquered the world and was still the standard for soft drinks more than forty years after it was introduced. But simply by describing this icon, you enumerate its shortcomings. It was thick walled and heavy, thus expensive to handle, and it did not hold very much Coke. The Coke bottle defined a quantum of soft-drink consumption, and it was quite small.

Other soft-drink bottlers offered larger containers, of course, and Pepsi had established itself as a rival by offering "twice as much for a nickel." During the 1950s, Coke had offered a much larger twenty-eight-ounce bottle in the familiar Coke-bottle shape, but it was hard to handle for consumers, as well as for distributors and retailers. Besides, it looked funny. Many soft-drink companies, including Coke, offered twelve-ounce bottles, but they were also difficult to handle. And because all soft drinks were returned for refill, empties took up storage space unproductively. In some important markets — Chicago for one — supermarkets stocked a variety of soft drinks only during the warm months.

Putting soft drinks in cans was an obvious solution. Cans strong enough to withstand the high internal pressure caused by beer had been introduced in 1935. Canned beer had been found to occupy 64 percent less warehouse space than same quantity of bottled beer, and the shipping weight was less than half as much. And while many drinkers continued to prefer bottled beer, brewers were able to pass the economies of canned beer on to their customers and to increase the quantities sold. The change was particularly advantageous for brewers who marketed nationally or within a large region, because money saved from less expensive distribution could be channeled into advertising that small local brewers could not afford.

Soft drinks presented greater problems of pressurization than beer, and the first soft-drink can was not introduced until 1953. In 1962 a canning industry publication was still touting canned soft drinks as a promising future use for their products. Many leading soft-drink producers — Coca-Cola most of all — were completely identified with their bottles. These bottles, in turn, determined such other matters as the configuration of vending machines and the nature of supermarkets' display. But pressure from supermarkets and competitors was building for Coke to make the move to cans. Pepsi, its chief competitor, was marketing itself as the "up-to-date" soft drink "for those who think young," and had little to lose by changing.

The eventual triumph of the soda can depended on two important technological innovations. The first was the introduction of the all-aluminum beer can, first promoted by the Hawaii Brewing Company as "the shiny steiny" in 1958. Tinplated steel cans are made of three pieces: top, bottom, and vertically seamed side. With aluminum cans, the bottom and sides are made in a single process, and the sides are seamless. Only the top must be attached. And in 1963 the aluminum tab top was introduced, which meant for the first time that the cans required no opener. The tab design was troublesome at first, because the pull-ring sometimes broke off, leaving the contents inaccessible. Moreover, the tabs created a serious litter problem that also posed a threat to fish and wildlife. The rapidity with which these problems emerged was itself testimony to enormous consumer acceptance and enthusiasm for this sort of container. In only a few years, the design was improved so that the tab would stay with the can. For several years, such aluminum tops were used on steel cans, but now, virtually all beer and soft-drink cans are made entirely of aluminum. And the smallest standard serving of soft drinks has nearly doubled, to twelve ounces.

But this explanation does not completely settle the matter of how Coca-Cola managed, during the 1970s throughout most of the country, to virtually abandon one of the greatest, most beloved packages ever invented, with hardly anyone noticing. One answer is that Coca-Cola never admitted it. The Coke bottle remained part of the company's imagery, and it was never wholly unavailable, merely uncommon.

Another possibility is that a subtler aspect of Coke's identity helped keep the memory of the bottle alive. That is the sinuous form, known as the dynamic ribbon, that has appeared, horizontally or vertically, on all Coke packaging since the 1970s and is even on the new neoclassic contour plastic bottles. It's not a literal repetition of the bottle's famous silhouette, and it's unlikely that anyone would look at this wavelike shape on a twelve-ounce can or two-liter plastic bottle and consciously think of a Coke bottle. But it gives the more conventional package an appearance of curviness that seems natural to the brand and subliminally keeps the memory of the bottle alive.

In 1993 the bottle was reintroduced as part of the company's logo, along with the tag line "Always Coca-Cola." (This may have been a delayed response to the debacle of the new-taste, youth-oriented "New Coke," one of the greatest miscalculations in the history of marketing.) The advertising campaign indicates that the company believes that this package, though it has been little used by most of its bottlers for three decades, is still a potent marketing tool. Coke has followed up with a somewhat clunky new version of the contour bottle, executed in PET plastic and including a bright new belt that adds color and graphic sizzle to the container. A label had not really been needed before, but now that Coke can be classic, diet, caffeine-free, or cherry, there needs to be more information on the front.

Liberation from the familiar bottle during the 1970s left the company free to try other forms of new containers. Coke was a leader in going to the two-liter PET plastic container, which is now the standard size for household use of soft drinks. This container proved to be even easier to handle than aluminum cans, and it does not generate the litter that smaller containers do, though aluminum cans are far more attractive to recyclers. The introduction of the large package coincided with an increase in soft-drink consumption. Per capita soda drinking rose from about thirty gallons per capita in 1972 to forty-eight gallons in 1992, an increase of 60 percent. The industry followed the two-liter PET bottle with the three-liter PET bottle, and Pepsi has been trying single-use containers for young people who are accustomed to drinking a liter of soft drinks at a time.

Coca-Cola's story is a late and dramatic example of a phenomenon that appeared in the second phase of the post–World War II boom, the decline of the monolithic package. Products began to become available not only in a range of sizes, but even a variety of packages and in different forms, such as cream, spray, and roll-on deodorants. Manufacturers recognized that different people had different preferences about the appearance of a product and ways of using it. But rather than create competing products, as Procter & Gamble had, they began to market products in different forms and packages, all of which could be advertised together. In a sense, this strategy recognized the intimate, physical relationship people have with a product they use, while attempting to transcend it through advertising. If a blue soap was making inroads in the market, a historically gold soap could bring out a blue version, complete with blue packaging.

Such product and packaging adaptation was particularly evident in products that were left on display during use. As part of a general trend toward the elaboration of private life during the late 1950s, there was a proliferation of decorative styles. Moreover, a modern appearance began to be seen as cold and impersonal, even in the kitchen and bathroom, where hard white was giving way to soft colors and woodier, more "natural" decorating approaches. Supermarkets began to offer such items as Early American– or Pennsylvania Dutch–pattern paper towels and disposable drinking cups keyed to match any decor. Surveys of consumer preferences made it possible for sellers of household products to understand how shoppers thought about their houses, and how they wanted both products and packages to fit in.

A casualty of this trend was one of the great American packages, the blue-and-white Moderne-style Kleenex-tissue box. Kleenex was truly a product a package made. It first came on the market in 1924, but made relatively little impact until the introduction, in 1930, of the "Serv-a-Tissue" box. This consisted of two separate, interfolded roles of tissue. When one double sheet was pulled through the dispenser slot at the top of the box, the next one popped up ready for use. This package, with its constant reminder of how easy it is to use the product, established both a brand and a product category. But because the pop-up feature was unique to Kleenex, it was important that the package have a memorable appearance to distinguish it from competitors. Moreover, even though the majority of sheets of Kleenex were used by people to blow their noses, this is not an activity anyone wants to call attention to. Thus, the classic Kleenex box sought a look that was as modern as its dispensing system, but also very elegant. It took its visual cues from expensive cosmetics packaging and from the tall, thin lettering often used in department stores. Its blue and white communicated a combination of softness and purity.

It is true that by the early 1960s the Kleenex box was a Fred-and-Ginger artifact in a "Leave It to Beaver" world. But Kimberly-Clark, the makers of Kleenex, did not simply update a package everybody knew. Instead, they made it disappear. The boxes that replaced it were carefully calculated to fit into different sorts of popular decor and color schemes, and these have been regularly updated. But most of their identity as Kleenex boxes disappears as soon as the box is opened. This change of design approach was, to be sure, taken in order to be competitive with other brands that had adopted this household chameleon approach. Tissue is no longer a modern miracle. Rather it is something people expect, and something they would like to have blend in with its surroundings. Kleenex did not make the efforts Coca-Cola did to keep its old package alive. Now Kleenex has little visual identity.

There is probably no package form more characteristic of the post–World War II era than the aerosol can. It was, in fact, a container that came directly out of the war, having made its first appearance in the South Pacific in the form of the "bug bomb." This was a single-use container that, when activated, sprayed all the insecticide it contained over a room-sized area, asphyxiating the local fauna and causing considerable discomfort for nearby humans as well. It proved to be a useful tool for fighting a war on tropical islands, and with the introduction of new controllable spray tops and closures the aerosol can became an immediate postwar success.

Like most new varieties of packages, the aerosol spawned new classes of products. Spraying was a novel way to access a product. Indeed, it was pretty close to a magic wand, a push-button mist of power.

All sorts of cleaning products offered the simplicity of spraying the dirt and wiping it clean. Spray furniture polish proposed that two boring tasks — dusting and polishing — could be combined into a single, satisfying act. Spray dessert toppings promised the indulgence of whipped cream, without the penance of whipping. You could spray starch precisely where you needed it. You could avoid embarrassment by creating a cloud under your armpits, while avoiding any damp spots. First-aid antiseptic healed without a potentially germ-laden helping hand. The rituals and most of the accoutrements of shaving could be eliminated with instant lather. Hairspray made it possible for your coiffure to assume positions unknown in nature. Attractive canapés became a snap with aerosol cheese. Spray garlic eliminated crushing and peeling. Spray tenderizer ennobled cheaper cuts of meat. And, of course, aerosol insecticide, which started it all, killed bugs dead.

Nowadays, aerosol containers tend to be viewed as vaguely disreputable packages, whose propellants have played a role in the depletion of the earth's protective ozone layer and whose contents have produced such visual blights as spray-painted urban graffiti and the 1960s beehive hairdo. Although the ozone-damaging chloroflu-orocarbons have long since been removed from aerosol containers, manufacturers have discovered that a substantial percentage of shoppers feel more comfortable with cleaning and grooming prod-ucts that use a conventional pump to spray rather than an aerosol. Indeed, even when aerosols were at their peak there was something disquieting about them. Their pressurized contents seemed explo-sions waiting to happen, and besides, there always seemed to be something left over in the bottom of the container that was inac-cessible and thus wasted. There were some products — toothpaste, for example — that were repeatedly introduced in aerosols without success, because they did not offer enough additional convenience to offset their greater cost and presumed waste. A pump bottle is considerably more expensive to produce and uses more materials than an aerosol, though it can be refilled, while an aerosol cannot.

Consumers are, in fact, rather selective about which aero-sols they spurn. There are still quite a lot of them on the shelves, and in some product categories, such as shaving cream, aerosols account for nearly the entire product category. Only in recent years has a major manufacturer, Gillette, introduced a nonaerosol shaving cream. It is more expensive and less convenient than the aerosol product.

The reaction against aerosols is not yet so strong that people will desert an effective aerosol for a nonaerosol that doesn't really work. Besides, the persistence of aerosol cheese suggests that there are a lot of people who just like to press the button and spray stuff.

If it seems that the golden age of the aerosol is past, another postwar packaging technology has only recently begun to come into its own in the United States. This is aseptic packaging, something that has been familiar in Europe since the early 1960s, but which was slow to win acceptance by manufacturers, government regula-tors, and consumers in the United States. It is, essentially, a variation on the canning process. But while canning involves sterilization of the can and its contents together, aseptic packaging sterilizes the containers separately from their contents and fills them in a sterile environment. By allowing different methods to be used for container and contents, it allows the packaging of foods with ingredients that can't survive the rigors of the retort, especially milk and eggs. Moreover, it allows the use of containers made of plastic or plastic layered with paper or foil, which are lighter to ship and easier to handle.

Aseptic packaging is, in fact, the fulfillment of Gail Borden's dream. His canned evaporated milk was a more reliable product than that available on the streets of mid-nineteenth-century New York. But it was also a different product, not like real milk. It was largely supplanted with refrigerated trucks and refrigerators in grocery stores and homes, which made the delivery and use of chilled fresh milk possible. Aseptic packaging allows fresh pasteurized milk to be kept for long periods in sealed cartons without refrigeration. It need not be chilled until after the container is opened. Thus, it frees milk producers from the need to provide chilled trucks and warehousing and retailers from having expensive in-store coolers. After nearly two decades of use in Europe, this container won approval of U.S. authorities. It was unveiled at the world's fair in Knoxville in 1982 (which may have been part of the problem) and evoked little response. Another major marketing effort, for a brand called Parmalat, was launched in 1994.

The reason that most Americans do not know that milk needn't be sold cold is that no major dairy producer or retailer, potentially the chief beneficiaries of this technology, has ever pushed for its acceptance. Failure to accept an improved technology generates far less of a paper trail than the pursuit of an innovation, and thus is harder to explain.

In this case, though, the answer is likely to be found in supermarket atmospherics. Since the early 1980s, the key word in food retailing has been "fresh." The center of the supermarket, with its aisles full of boxes, bottles, and cans, is essential but dull and generally not too profitable. A supermarket wins its identity and makes its money around the edges — in produce, meat, fish, and delicatessen and precooked foods. The dairy cases, which are also on the periphery, are not as high profile as those departments, but still more desirable than the packaged center. Many products, such as pasta, salad dressings, sauces, and juices, have migrated from the dull center to the chilled edge, increasing prices and profit margins in the process. Milk is the most important generator of traffic in this part of the store. Why take a product deeply associated with the very idea of freshness and exile it among packages that bear labels that say they are best used sometime in the next decade? The dairy industry is very good at delivering the milk fresh to the stores, which sell it fresh to consumers, who use it before it spoils. It's really not too surprising that the aseptic milk carton is viewed as the solution to a problem that doesn't exist.

Nevertheless, the aseptic package and even its milk-preserving properties have been embraced in other sorts of products. The most successful is the multipart snack pack of pudding or applesauce, from which individual servings can be torn off and packed in a lunch box to provide dessert. Another example is the single-serving juice box. This container is controversial because it is the kind of small, casually used package that often turns up in litter and because its multilayer construction makes it difficult to recycle. It does, however, allow the delivery of a product less altered by processing, easier to carry than a canned drink, and less demanding in handling than a chilled one. But when aseptic juice boxes have been offered in larger sizes, American consumers have tended to reject them, choosing chilled or frozen juices instead.

It is easy to imagine that aseptic packaging would have been embraced in America during the 1920s, when consumers were demanding products in small, manageable quantities, when storage space was limited, and refrigeration not universal. This describes the conditions in Europe that led to aseptic packaging's invention and success. But in the United States, manufacturers and retailers assume that the only small, unrefrigerated space in American life is the lunch box.

The lack of acceptance of the aseptic milk package points up the reality that packaging does not end at the container's edge.

The refrigerator case is part of the packaging for milk, as is the entire supermarket. The supermarket is standardized and carefully considered — a set of sensations, expectations, and satisfactions rigorously tailored to promote customer loyalty and a healthy bottom line. Supermarket chains are identified with particular colors, graphics, and methods of display. The look of the store is echoed in the look of the store's private-label products, and sometimes the same designer is responsible for both the appearance of the store and the packaging. Milton Glaser's late 1970s makeover of the New York–area Grand Union chain was a particularly dramatic example of this practice, as is Selame Design's recent work for the Boston-area Purity chain. Supermarkets began as neutral settings for expressively packaged products, but in order to compete they have had to evolve into packages themselves.

Packaged environments, building-sized and larger, constitute a very large amount of the post–World War II, automobile-centered environment. In the last chapter, we saw the origins of such large-scale packaging in the rise of restaurant chains and standardized gasoline stations. There was also some standardization of drugstores, tobacco stores, and five-and-ten-cent stores. The development of a new kind of suburban landscape after World War II, and the relocation of a large portion of the population into unfamiliar territory, magnified the allure of beacons of familiarity. The automobile allowed people to do business in a far wider geographic area, even as it reduced the sociability inherent in pedestrian street life. By reducing exposure to one's neighbors, and increasing the amount of business done with strangers, the automobile encouraged the proliferation of the kind of packaged environments that constitute so much of the American landscape today.

This was already becoming apparent when, in 1943, Deskey was given the assignment of creating prototype facades and interiors for combination bowling alleys and pool halls. His clients, the Brunswick-Balke-Collendar Company of Chicago, wanted a postwar look that would show off its bowling and billiard equipment in a wholesome, family-oriented, postwar way.

Both pool and billiards had lingering unsavory associations for many people. Deskey proposed that Brunswick could expand its market by taking control of the image of the games and the experience of playing them. Strong design and strong marketing of the Brunswick image would leave the operator of the bowling alley or billiard parlor little choice but to live up to the standards the company set. Brunswick was looking past its ostensible customer, the operator of the facility that bought the equipment, and toward its ultimate user.

"Such a building exterior should be done extremely well and only after much thought and analysis," Deskey wrote in a report to his client. "It can be made as clean-cut and fine as the best in modern packaging. The design of such an exterior is, in fact, a PACKAGING JOB — and a most important one. It should be subjected to the same research as the package that is intended for national merchandising." Architecture or interior decoration, which involve the design of a particular place, had to yield to the higher authority of packaging, whose goal is to be instantly expressive wherever it is. Deskey's recommendations were accepted and helped Brunswick promote a bowling boom during the 1950s and 1960s.

The development of the contemporary packaged landscape — raffish drive-ins, Googie coffee shops, populuxe bowling alleys, the evolution of McDonald's from golden arches to mansard roofs, the invention of the mall, the rise of the theme park — is by now a familiar set of stories that have been told repeatedly and well elsewhere.

What's important to recognize is that fast-food and motel chains are not *like* packages, but that they *are* packages. The goals of packaged places and experiences are exactly the same as those of more conventional packages. They allow corporations to have complete control over their messages, up to and following the moment of consumption. And they provide consumers with uniform expectations that can be fulfilled. Quality — as industrial engineers define it — is how well a product performs measured against what the user anticipates. Standards assumed for a twenty-dollar watch or a McDonald's Happy Meal are different from those for a Rolex or dinner in a restaurant that gets three stars from Michelin. It's possible that the cheaper items might even be found to be of higher quality. If you can calibrate expectations and ensure delivery at those standards, you have a high-quality operation. Packaging, whether it is a bag or a building, is a tool to control expectations and to assure that the product can be delivered in accordance with those expectations.

"The best surprise is no surprise," said the famous slogan for Holiday Inns. Just as packages promise that the same standards will be maintained from can to can and box to box, so did Holiday Inns and other chain hostelries and restaurants promise that guests would find exactly what they expected. Today, such building-sized packages cluster at every freeway interchange and stand with their associated parking lots along highway strips. They are found at the centers of cities, where strangers cross paths, in urban neighborhoods, and small towns. Kentucky Fried Chicken is no longer available in Tehran, but you can find it just off Tienanmen Square in Beijing, sporting the same color scheme, the same uniforms, and the same cholesterol count as in Tucson or Altoona.

The law makes little distinction between packages you can walk into and those you can carry in your hand. The look and feel of a chain of supermarkets, restaurants, transmission centers, theme parks, or dental offices are protected on the same basis as the design of a bottle, tube, or pouch. Both are, the U.S. Supreme Court has declared, forms of trade dress and subject to the same protections. In fact, the requirements for protecting the design of a packaged environment appear, at the moment, to be somewhat easier to meet than those for a packaged product. The packaged product has to have achieved large-scale distribution, while the place has only to be both distinctive and potentially replicable.

Environmental packaging has become a particularly important component of the tourism industry. While it is true that travelers have long made pilgrimages to "the sights" — the Eiffel Tower, the Trevi Fountain, the pyramids — tourism is increasingly becoming a matter of validating images people have seen in the media. Florida's Walt Disney World is now the largest tourist attraction in the world. Its version of the Doges' Palace is, for some, more satisfying than the real one in Venice. It is visually superior because there are artists on staff who keep its patina absolutely perfect, and it is managed entirely for the satisfaction of the tourist, who is spared any contact with real Italians or the floodwaters of the Venetian lagoon. The real Venice is virtually a theme park by now, but it still has trouble competing with the fantastic.

Theme parks that replicate such European attractions as the Leaning Tower of Pisa and the Eiffel Tower have recently become popular in Asia. One that opened in China in 1993 was marketed as a substitute for the trips that most Chinese will never be able to take. Meanwhile, many tourists who visit art treasures and historic landmarks in China are able to see only replicas. Sites that have been packaged for tourists are easier to visit, and soon there will be little else to see.

The city of Philadelphia pondered whether to place the statue of the fictional prizefighter Rocky in the location where it was seen in the movies in which the character was featured. The reason was that visitors complained that the city ought to be as it is in the movies. At the same time, it has been grappling with packaging the Liberty Bell and Independence Hall so that they will better accommodate the cycle of anticipation, fulfillment, and photography that now defines the tourist experience.

One interesting form of packaged environment is the "festival marketplace," such as Boston's Faneuil Hall Marketplace, Baltimore's Harborplace, or New York's South Street Seaport. One feature of these combination theme park–malls is stalls where fancy fruits, cheese, and other indulgent food is sold, unpackaged, by someone across a counter. But buying in this way is conceived as part of the fun, not as a means of sustenance. The person behind the counter is not seen as a real butcher or baker but as part of a total entertainment experience.

At Disney's theme parks, everyone who is seen by the public — whether as a human being or as an oversized duck in a trouserless sailor suit — is classified as a member of "the cast." The ability to convince its workers that they are part of an ongoing performance has generally worked well for Disney. It encourages workers to take direction, to do what's in the script rather than what they feel or what they think makes sense. It would probably be a good deal harder to convince the minimum-wage teenagers staffing the Taco Bell that they are pursuing a life in the theater. Yet just such a surrender of personal judgment is required for the proper management of such packaged businesses.

While conventional packaging maintains control of the message by essentially removing people from the selling process, McDonald's, Holiday Inns, and other chains make people part of the package. Fast-food employees' tasks are carefully designed, along with the equipment and the containers in which the food is dispensed, to provide a uniform result, with a minimum of errors. McDonald's French fries are designed to be dispensed in such a way that their cardboard container appears slightly overloaded. The scoop used to fill them holds precisely the right quantity of French fries to achieve this overloaded appearance, without of course actually overloading, which would lead to spillage and lower profits. This is a very sophisticated form of package. Humans are involved, but their role is not to sell or even to serve. The design challenge is to give them little or no opportunity to do harm. Some chain organizations provide scripts from which little deviation is tolerated. "Discretion," wrote Harvard Business School marketing guru Theodore Levitt in a 1972 article on McDonald's, "is the enemy of order, standardization and quality."

McDonald's and other chains represent the industrialization of businesses that were once thought of as service-oriented and people-centered. The polished impersonality of the Automat was perhaps the first attempt to promise and deliver the predictability and economy routinely achieved by mass-production processes. Significantly, aside from a woman — known as a nickel-thrower — who made change, the Automat kept the people who made it work out of sight. This proved to be an expensive and impractical pretense. In most chain operations, people are visible. They simply are not authorized to behave as people.

"While it may pain and offend us to say so," Levitt wrote, "thinking in humanistic rather than technocratic terms ensures that the service sector of the modern economy will be forever inefficient and that our satisfactions will be forever marginal." Americans have never been entirely comfortable with the concept of service in the first place. Now, Levitt argued, person-to-person transactions have simply become too expensive, like handcraftsmanship. And when such transactions cannot be replaced by conventional packaging and pure self-service, or by such machine services as automatic teller or ticket-dispensing machines, human beings can simply become part of the package.

7 *The art of the package*

During 1993, every American food package was rede-signed. Package designers everywhere were busy and prosperous. Giant food processors were forced to grapple with the managerial nightmare of dealing with hundreds or thousands of new designs all at one time. The new packages began arriving in the stores in the autumn of 1993, accelerating to meet a May 1994 deadline.

Such a large-scale simultaneous change of the look of pack-aging — forced by new Food and Drug Administration labeling reg-ulations — was unprecedented. It's also likely that most people scarcely noticed it, and that those who did perceived it not as a change in package design but as an augmentation of the informa-tion on the label.

The space on a package is finite. The impact of the newly required, more extensive nutrition-information box was a bit like replacing a piece of furniture in a room with another one half again as large. Usually you can fit it in, but everything else in the room has to move, and perhaps a few things have to be removed to make space. There are a few package types that offer so little space — single-serving juice cans, for instance — that manufacturers sought and received waivers from the FDA to modify the nutrition label slightly. In most cases, it was possible to make subtle changes that preserved the familiarity and the appeal of the package while accommodating the new requirements. Still, the success or failure of a package design often lies in just such subtleties. Thus, package revisions that are intended not to be consciously noticed can sometimes require as much design skill as those that are. And while a few companies took advantage of the necessity of redesign to give their products an updated look, most food processors did not want shoppers to perceive that there had been any change.

In fact, small changes are being made to even the most familiar of packages all the time. A candy bar shrinks, forcing the redesign of the wrapper. A toothpaste adds a new feature or is repackaged in a box with slightly different proportions. A piece of copy on the package is shown to be ineffective, or the typeface begins to look old-fashioned.

Not long ago, the Campbell's soup can, which had been printed in three colors for eight decades, began to run new recipes on the backs of its cans, illustrated with full-color photographs, a quiet revolution. The purpose was not to change the appearance of the package on the store shelves, but to offer people who have already purchased the soup an appetizing new way to use it. The company's research had found that nearly everyone had a can or two of soup stockpiled on their shelves, but many people seemed to be saving it for an emergency. In most fundamental ways, the Campbell's can has been unchanged for nearly a century, yet it has been redesigned constantly.

In 1994 Campbell's U.S. can underwent a major change. At the bottom of the face of the red-and-white can there is now an appetizing picture of the soup to be found within. This diminishes the purity of one of the classic package designs, but it may have been something Campbell needed to do. The old cans established the company's identity and represented nearly a century of people's faith in the cans. But the company's challenge was not to get people to trust the soup but to induce them to actually eat it. Thus, a commercial icon had to be sacrificed.

*   *   *

There is a temptation to dismiss frequent, minor graphic changes in the presentation of a product as insignificant. It is likely that few people ever notice them consciously, but there is plenty of evidence that shoppers are capable of responding to even the subtlest kind of modification. Sometimes even designers are amazed by how much of a difference a small change can make.

In the mid-1960s, the veteran package designer Irv Koons was asked by Fred Mueller, the president of the company that made Mueller's spaghetti and macaroni, to do what he called a radical redesign of the pasta packages. The design dated from the company's beginning in the nineteenth century, when it delivered spaghetti in blue-and-white wrappers to saloons and restaurants. Red lettering was eventually added, to assert proprietorship over the product and to give it an American feeling. Like many early packages, this one evolved through a series of incidental decisions rather than as a conscious design. For example, the typeface used for the company's name was probably something that the printer had available rather than an aesthetic decision. Later, the company had it registered as its trademark. At the time Mueller came to Koons, the company was still first in its product category, but it was beginning to lose ground. That's the time that companies begin to think about changing their packages.

"They wanted a complete redesign that would turn the situation around," recalled Koons, "but we needed to keep the package's blue top and bottom, with a white stripe through the center and the name in the same old-fashioned typeface." Koons's approach was to slightly enlarge the cellophane window through which people could see the product, and to slightly adjust the configuration of the package so that all of the window was bounded by white. A layman viewing the revised design might wonder why a company would go to the trouble of making a change that's so insignificant. Koons said Mueller's reaction was fear that the change was too revolutionary. "I told him that he had a beautiful product that should be visible and framed in white so that people could see the golden egg color," he said. "I was looking for a feeling of cleanliness, a more contemporary look." Sales, which had been stagnant, shot up about 8 percent the year the slightly new look was introduced — and as much as 28 percent for certain products in the line — and 10 percent overall the following year. This happened without any change in the advertising budget or the theme of the campaign. "The package must have been what made the difference," Koons said, "but I wouldn't have believed myself that such a small change would do that."

This positive experience led to two subsequent, more substantive redesigns by Koons, one of which introduced what was, at the time, a newly designed, up-to-date typeface. Throughout this entire process of redesign, the Mueller's buyer probably thought of the package as one that never changed. Yet the sales figures reflected each seemingly imperceptible change.

A loss of sales or of market share is the most common reason for redesigning a package. A redesign can also be spurred by a major change or an anticipated move by a competitor or, in the case of some large companies that do constant market research, a change in the way those surveyed view the product.

Another reason almost always cited by package designers is the brand manager system. Brand managers are typically young men or women in their twenties and thirties assigned to manage a particular brand for a few years. "A major corporation wouldn't let a really junior little accountant make a big decision affecting the expenses and revenues of a company for years to come," said Elinor Selame, president of the design firm Brand Equity in an interview. "Yet, they're willing to let brand managers play with the image of their products." Selame, like many of her colleagues, feels that brand managers often make counterproductive changes to package designs, simply in order to show their bosses that they are making a difference. Such nervousness provides more work for package designers, but it can diminish the power of packages to sell the products. Not infrequently, package designers are called upon to clean up a family of product lines after a succession of competing brand managers has taken the brand's image in so many different directions that advertising and marketing efforts have become inefficient.

Those who do market research are less inclined to fault brand managers, because their changes are often made in response to research findings. The conflict is essentially whether packaging should remain consistent, responding only gradually to social and economic trends, or whether it should be constantly sensitive to them.

For most established products, the goal of packaging re-design is to look better, but not necessarily very different. Every package design has what marketers and designers call equity, a com-bination of colors, typefaces, geometric forms, and general design approaches that are identified with that product. Some of them are extremely simple. Kodak, for example, is indelibly associated with a particular shade of yellow, and little else. Coca-Cola is an unmistak-able red. Others are a bit more complicated. Pepsi is associated with a red-white-and-blue circle, which thirty years ago represented a bottle cap but has evolved into a bold, mandala-like form that the company associates with the world. And studies done for the com-pany show that consumers expect the look of Pepsi to change. Keep-ing an up-to-date appearance is part of Pepsi's equity. It can't afford not to seem to change with the times, while other brands take risks when they do so.

Researchers hired by Campbell discovered that people were aware that its soup cans are red and white, but they couldn't even remember whether the red is at the top of the can or at the bottom. Although consumers would very likely notice if the colors were re-versed, or the can's gold accents were dropped, the designers were able to conclude that they had considerable latitude to rethink one of the world's most familiar packages. (The red is at the top, by the way.)

The goal of redesigning a package is, after all, not to alien-ate those who already purchase and like the product, but to broaden the market for it. It is possible, indeed essential, for even drastically changed designs to go relatively unnoticed by faithful buyers be-cause the revised design is still true to the personality of the product its customers know and like. The point is to change at least enough to keep up with the changing attitudes and tastes of those customers and, ideally, win new buyers. Customers may not be specifically con-scious of a desire to see that the appearance of their cottage cheese tub reflects their evolving lifestyles and visual preferences. But just as people expect their friends to grow and change with themselves and their times, they also expect packages to occupy a particular niche in their lives. Some products are valued for their stability; we don't expect our mothers to get their fashion cues from MTV. But people stop buying others, often without consciously thinking about it, simply because they are no longer consonant with the way they live. That cottage cheese container becomes a friend you just don't see as often as you used to.

Thus, packages change with the times. But the way they do so is very subtle. While it is possible to look at a poster or a building and classify it roughly by date on the basis of its visual style, this can be far more difficult with a package. Most new packages are designed to last for a while, and there are good reasons to expunge from the design anything too markedly trendy. Moreover, the design trends to which packages respond are not purely high-style phenomena. The manufacturer of a facial tissue, for example, is likely to be much more interested in trends in wallpaper, bathroom fixtures, and kitchen and bedroom paint colors — the environment in which the package will be used — than in what's hot in the design world of New York, Los Angeles, or Milan.

Such changes in people's personal environments do not show up in conventional design histories, and it is difficult to pigeonhole many packages into recognized stylistic categories. But often these seemingly hidden design waves have enormous social repercussions. For example, in the mid-1950s, sales of colonial-style furniture began to rise, while those of modern and contemporary styles started to fall. A bit later, westerns began to monopolize the television schedule. Louis Cheskin, the pioneer psychological marketing researcher, began to hear consumers speak of a desire to somehow reconcile newness with tradition. His advice led to a package design for Duncan Hines cake mixes that featured the brand name in a "colonial-style" frame, but that used bright, modern colors and generally clean graphics. This expression of convenient tradition was a departure from most packages at the time, which expressed convenience as a result of modernity.

A few years later John F. Kennedy ran for the presidency

under the slogan of the "New Frontier," which cartoonists represented both in terms of rocket ships and covered wagons. This popular rejection of the look of modernity, while embracing its achievements, was a major change in the history of American design and culture. It happened completely beneath the radar of the taste-makers and only appeared in high-style expressions more than a decade later under the guise of postmodernism. This is a style that began at home, and packaging is a good place to look for such "hidden" trends.

Still, the need to preserve the equity of a package is the fundamental reason for the understated way in which packages respond to style. Packages can't afford to be fashion victims. Some brands have revived themselves through total transformations, but such cases are fewer than those for whom unrecognizably new looks proved disastrous. Packages can't change as fast as styles do. Rather, the most successful of them react to stylistic change the way most people react to fashion in clothing. They look up-to-date by adopting those bits and pieces of the new that seem most adapted to their own personality. And just as we expect certain people to look more stylish than others, established brands have their own distinct packaging personalities. Such gradual, organic adaptation used to be characteristic of the building, toolmaking, and decorating crafts that shaped towns and houses. Packaging, with its severe limits on individual expression, is one of the few design disciplines that values continuity over bold new statements. Packages are usually more reassuring than buildings or appliances.

Package design has two components, graphics and structure. Most package redesign concerns graphics — what's on the container. Often, especially for brands that include many different products sold in a variety of container types, the brand equity resides in a purely graphic concept, a combination of colors, typeface, logos, and overall design style.

Structure — the physical form of the container — is nevertheless extremely important. It is the profile that catches the eye in the store; it is how the package feels in the hand and how it opens and dispenses. And for some products it is an extremely important part of the brand's identity.

As we have seen, innovative package structures, such as Mennen's secure, integral-shaker baby-powder tin and the pop-up Kleenex, have helped establish important brands and whole product segments. The Coke bottle's structure has survived in the consciousness even as most of the beverage comes in other sorts of containers. Marlboro's flip-top box gave the product a special feature that helped it build a lead over its competitors. The distinctive, egg-shaped L'eggs pantyhose package, along with its in-store display, established an industry, and arguably a lifestyle.

The proliferation of products sold under a single brand, along with changes in packaging technology, has led in recent decades toward a devaluation of structure as an important part of package design. It is argued that while structural innovations have given particular brands significant head starts, the general tendency is for packages to become generic to a particular product category. Even the L'eggs package has, in response to environmental consciousness and the desire for cost reduction, moved to a more conventional container.

Colgate toothpaste gets no advantage today from being the first major brand to be sold in tubes. And each major advance in tube design, materials, and manufacture — along with some innovations that had little usefulness or customer appeal — has generally been taken up by all the competing brands of toothpaste. After its use of the tube was universally copied, Colgate tried to maintain its edge by introducing a tube in which the dentifrice came out as a flat ribbon rather than a cylindrical extrusion. This was ultimately abandoned. Several alternative rigid-walled pump and aerosol containers have been introduced for toothpaste, along with devices to dispense striped toothpaste. But the gradually improved tube — originally metal, now plastic — remains the standard in its category. The most successful current alternative is the tube that stands on its head. It would be very risky for anyone who makes toothpaste to foresake the tube entirely. And Colgate's equity lies in its use of red and white and in the typeface in which its name is written.

The tube's status as the expected package for toothpaste has forced makers of other kinds of products found in tubes to modify the form to meet their needs. It's difficult to know just how many people brushed their teeth with shampoo or rubbed toothpaste into their hair to make it stay in place. What's important is that people feared that they would, and by the mid-1960s this potential confusion was identified as a problem for products in tubes. Brylcreem, a men's hairdressing, responded by moving into a tube with dark colors, totally unassociated with toothpaste. Then changing hairstyles made aerosol hair preparations more desirable, while shampoo moved more and more into plastic squeeze bottles. Later, after hairspray had gone out of style, styling gels were introduced in tubes, but they tend to be shorter and wider, with a smoother texture than the tubes used for toothpaste. You don't even have to look at the package to know what kind of product it is; you can tell simply by picking it up. This ability to mold tubes into many distinctive shapes is but one result of the flexibility of plastics. When all tubes were metal, they were all more or less the same. The tube was a generic structure, but now it is becoming increasingly communicative.

There has been a renewed interest in distinctive package structures in recent years, however, due in large part to environmental concerns and corporate cost cutting. Increasingly, items such as deodorant, shampoo, and other cosmetic and pharmaceutical products are being sold in their primary containers rather than in paperboard boxes. These outer boxes were not a total waste. They made shipping and handling easier and protected the inner container. With the box removed, the new container must be redesigned so that it can be packed and shipped and handled repeatedly without any visible scars. Filling machinery, sorters, palletizers, and inventory control systems are hidden forces that shape packages.

The other reason to reshape this newly naked container is to compensate for the loss of the display value of the outer package. These boxes were, obviously, larger than the primary package and provided six surfaces for graphics, ingredients, warnings, and other information. Now the primary package must do more of the work of catching the eye and expressing the spirit of the product. What have been cylindrical bottles become flat to provide more label area. The graphics become bolder and the profile more distinctive. This new package may be lighter and contain fewer materials, but it is more aggressive visually.

For example, the Listerine bottle, whose equity lay in a generic glass bottle wrapped in a corrugated sleeve, has moved to a flat plastic bottle. The old package was aggressively anachronistic; it said the product was old-fashioned but good for you. The new package expresses the almost authoritarian character of the product through a distinctive custom shape and a stronger typeface. The means used to get the message of the product across has changed completely, but the message itself is much the same.

To a large extent, the makers of products do not control the pace or direction of technological improvement in packaging structures. Packages are manufactured by specialized companies, as is the machinery for filling and packing the packages. While package manufacturers frequently help companies develop a unique custom shape or texture for a particular product, their investment in research and innovation is understandably biased toward products that can be sold to a wide variety of customers. International diffusion of new packaging structures is close to instantaneous, though the United States tends to be a bit more reluctant to embrace new structures than is Japan or Western Europe. Even when new materials are introduced, as when a glass bottle is replaced by plastic, the tendency is to maintain the old shape, which was often an expression of the limitations of the old material.

The law has a significant impact on this situation. In the United States, the law considers package design a form of trade dress, which can be protected almost indefinitely as a form of copyright. There is one significant exception to this protection, however. Trade dress is a way for a business to protect its name and prevent others from benefiting from the efforts it has made. Trade dress cannot be useful. It cannot include a unique applicator, nor can a business protect as trade dress a shape that has been derived from a scientific study of the human hand and how it grasps a container and pours a particular liquid. If form follows function, the form cannot be protected as trade dress.

Although this doctrine seems a bit perverse, there are sound reasons for it. Useful inventions are protected by patent law, which provides for exclusive use or licensing rights for a much shorter duration. This seeks to balance the inventor's right to benefit financially from the advance with the interest of society as a whole in benefiting from technological progress. Trade dress, by contrast, is intended merely to protect the investment of those who seek to create a commercial identity from others who might try to copy that identity.

It is possible to patent an innovative package, as was done with Uneeda's In-Er Seal biscuit box. Package manufacturers are more likely to obtain such functional patents than are product manufacturers or package designers. (There is another kind of patent, for distinctive design, which many designers do obtain.) Moreover, innovations in structural packaging tend to be more important for their goals than for their means. There are a number of possible ways to incorporate a measuring device into a laundry detergent package or to reseal raisins. What's important is that the addition of such features increases the usefulness of packaging for the consumer, not the specific means by which it was achieved. Today, a detergent that does not have a measuring device is old-fashioned and deficient. And the box of raisins is probably stale. In any event, package designers and product manufacturers often assume that useful design features cannot be protected, if only because the same goal can be achieved in another way not protected by the patent.

In fact, many of the most important functional innovations in packaging structure have required several steps of refinement before they really lived up to their potential. The evolution of the no-opener beverage can — from tab top to removable ring top to captive-ring top — shows how a much-desired product had to respond to both functional and social problems before it achieved a universal, useful form. A similar evolution has been happening more recently with the resealable zipper pouch. This is a useful innovation that has been beset with design difficulties. In addition to the re-

185

closable seal, such packages have to have a conventional seal, to prevent spillage during shipping and to forestall tampering. The problem was that in breaking this irreversible seal, consumers often tended to tear the reclosable seal, rendering it useless. In 1993 this problem seemed to have been solved with the introduction of pouch packs whose outer seal can be broken by pulling a highly visible string, while the inner seal remains intact. There is still something slightly awkward about pulling the string, though, and it would not be surprising if a more intuitive solution eventually takes its place.

The resealable inner pouch used by some brands of cereal poses an even more irritating problem. The machinery used by the industry requires that the bag's security seal be placed below the resealable zipper. That means that much of the time, someone opening the package will tear a hole that makes the resealable feature useless. This problem has been evident for years, without any attempt to solve it. To do so would require a major new investment in packaging machinery, and perhaps a reconfiguration of the filling process. These factors create a bias against the introduction of radically new packaging structures, and, as in the case of the cereal boxes, product marketers sometimes take half measures that promise greater convenience but do not deliver. And many cereal buyers probably conclude that they, not the package structure, are at fault for their awkwardness.

Packaging is one of the very few areas of contemporary life in which people expect steady, inevitable progress. From the point of view of product and packaging manufacturers, this confidence in the perfectibility of packaging is most often expressed in a negative way. When, for instance, raisins were first introduced in a resealable container, only a handful of raisin obsessives cheered this advance in maintaining softness and freshness in the product. The manufacturers were far more likely to hear from those who had a problem opening the new closure without ruining its resealable properties. This conforms to the cardinal rule that people in general think about the package only when it causes them problems.

There are some qualities that people expect of packages that have, so far, proved to be at odds with one another. For example, for many products, especially over-the-counter pharmaceuticals,

186

surveys indicate that consumers feel that it is overwhelmingly important that packages should resist tampering and show clear evidence when tampering has occurred. They place a slightly lower priority on the product being easy to open and convenient to use. These goals are not wholly at odds, but they do require a lot of thought and ingenuity to reconcile, and despite many advances, most packages have probably not yet arrived at a wholly satisfactory solution.

The question is complicated by issues of liability for accidents, such as the ingestion of drugs by children. The solution was to come up with ways of opening the package that require instructions and are counterintuitive. Thus, child-resistant closures resist nearly everyone else as well, particularly older people who are heavy users of many of the medications sold in such containers. Many companies long took the position that all their products had to contain child-resistant caps, no matter who the market was, though the sale of clearly marked nonchildproof containers has recently been on the increase. Recently, Tylenol introduced a special package for people with arthritis or other hand problems. Its cap features a high, flat grip, which allows the user to get plenty of leverage, and also provides a hole in which a pencil can be inserted and then pushed or pulled, so no twisting is needed at all. This package is tamperproof, and someone who is severely impaired might need help in opening it for the first time, but thereafter the package will not get in the way. This new Tylenol package may presage a new wave of packages that strive to be more useful and convenient for an aging population.

As these examples suggest, what packagers view as very limited legal protection for useful improvements in structure does not necessarily prevent the introduction of more useful packaging. But what it does prevent is the identification of one brand with a package that makes it more useful. The amount of time that a manufacturer can offer a product in a functionally improved new package before the competition begins to match it is short. The sort of head start that Mennen baby powder and Kleenex once enjoyed is probably impossible today. There's still an advantage in being the first product in your market to go into a demonstrably more useful package. There's even some advantage in catching up with the rest, structurally. For example, Murphy's Oil Soap — a furniture-care product with an image of old-fashioned effectiveness — during the 1980s abandoned the cylindrical glass bottles in which it had always been sold. The larger sizes were given handles, and sales of these packages shot up. There was no new advertising or promotion, simply a perception by shoppers that the product would be easier to use.

The products that are most identified with distinctive packaging structures are those that people are likely to display, such as liquors and cosmetics. Structure is also important for some products that are sold in drugstores, because distinctive shapes are one way in which branded shampoos, cold remedies, and antacids can appear to embody better value than the drugstores' and retail chains' version of each major product, complete with copycat graphics. Most of these copycats appear in stock bottles, while the brand names use more sculptural and eye-catching custom containers. Shampoo packaging is always among the most inventive areas of structural packaging because it is used in an intimate way, it is closely related to fashion and lifestyle, and it is extremely important to embody what it offers in three-dimensional form that won't be threatened by graphic imitation.

Increasingly, not even the most blatantly functionless aspects of packaging structure are safe from imitation. For example, at least two large drugstore chains offer cold remedies in a rounded triangular bottle, clearly imitative of the peculiar, obviously nonfunctional package long used by Vicks Nyquil. Drugstore products seem to be a special case. In other product areas, manufacturers are likely to have challenged the look-alike graphics, let alone the imitation of distinctive shapes. But the makers of toiletries and over-the-counter medicines seem reluctant to go to court against their best customers.

There is, moreover, a sense in which copycat package design validates the superiority of the brand-name original. The store brand's graphics, which are similar to but not quite the same as the national brand, communicate that the product itself is striving to imitate an original, but isn't quite the same. Consumers are meant to be swayed by the big difference in price between the original and the knockoff. Whether or not they are depends on the shopper, the product, and a personal calculation. But the one thing that is not in doubt is that the national brand defines the standard of quality against which the store brand is measured.

In a sense, the recent proliferation of store brands that are packaged distinctively, with hardly any reference to the national brands' trade dress, poses far more of a threat. President's Choice, Master's Choice, and Wal-Mart's Sam's American Choice are, in a sense, national brands that are exclusive to different regional and national retailing chains. (There are styles in words, and for the last decade "choice" has been marketers' word of choice.)

These new-look store brands do not depend on advertising to win legitimacy in shoppers' minds. Rather, they depend on the marketing might of the retailers, and in particular they depend on packaging. Their rise during the early 1990s came at a time when television had fragmented into a multitude of niche channels, and fewer and fewer women were at home during the day to watch the commercials that have traditionally established national supermarket brands.

It also came at a time when supermarket scanners were giving retailers the upper hand in their constant struggle with manufacturers and wholesalers. Now the retailer knows who the customers are, what they buy, and how to maximize the return on each bit of shelf space. They have the information and confidence to remove underperforming national brands from the shelves. And they know that their edge over competing chains does not come from stocking Cheerios and Tide, but rather from the items they offer that others do not. The last time that retailers had such an advantageous position over the other players in the field was during the 1930s when A&P dominated the grocery business.

The design of the new private-label packages does have some similarities with the packages Egmont Arens designed for A&P, which elicited similar alarm from grocery manufacturers. The most important similarity is the approach — to create distinctive packages that try to look better and more desirable than the national brands. This is doubly threatening because it undermines the national brands as setters of standards, and it allows the chains to set their prices for private brands higher than before and thus realize far greater profits from them. Like Arens's designs, the new private-label packages have the appearance of premium products. The success of the new products indicates that they have met shoppers' expectations. Arens was the first and last designer to establish black as a color for mainstream grocery products — until President's Choice.

The black, gold, and white palette found on these packages is a slightly out-of-date look for premium products — more Reagan-era expensive opulence than nineties clarity, purity, naturalness, and intimacy. Such widespread availability of what were, only recently, the trappings of affluence seems a shrewd approach for a period of economic uncertainty. The black-and-gold box — punctuated with appetizing, full-color graphics that often imply upscale lifestyles; flattering copy; and script-style graphics — offers a strong voice of reassurance. It promises a way of maintaining your standard of living, even of being able to move up a little, without having to pay very much more than you can afford. Bad times in the economy might have pushed shoppers from national brands to private brands in any event, but the packaging made them feel good about it.

The packaging transformed an item that the shopper merely settled for, because of price, into an item that was desirable on its own terms. Thus, the product line was positioned to weather even the return of prosperity and self-indulgence. It was not merely a stopgap, but a basis for new loyalty. However, the packaging style, so suitable for a particular moment in history, might easily become dated. "These are, potentially at least, new megabrands," said Ronald Peterson, of the large package design firm Peterson & Blyth Associates. "They will have to behave like brands and continue to keep their packaging up to date so that it offers what consumers are looking for." Peterson said that the number of new products being advertised without television advertising is steadily increasing, which makes package design, along with coupon promotion, the principal ways in which a new product gets noticed. The success of President's

Choice and the other new premium private brands indicates that retailers are aware of the power of package design.

Almost from the moment that modern packaging first became widespread in the late nineteenth century, artists have been interested in it. Only six years after the Bass ale red triangle became the first trademark registered in Great Britain, in Manet's monumental 1882 painting *A Bar at the Folies-Bergère* a Bass bottle and logo are clearly evident. The bottle does not occupy a particularly prominent place in this classic depiction of alienation in a crowd, but it does catch the eye. Psychological studies have shown that a triangle is an attention-getting shape, red is an attention-getting color, and when you put them together, they border on being irritating. Although the bottle appears in the painting because you would find it in a bar, the effort it makes to be visible seems to comment on the crush of people and personalities that the painting depicts.

Liquor bottles played an important role in the early Cubist works of Picasso and Braque, and in the work of many other early modern painters. One reason is their shape, which often fuses the pure geometric form of a cylinder with complex curves. But that explanation does not account for the frequent depiction of labels. You could probably mount a very respectable exhibition of paintings that depict Pernod bottles. Insistent commercial expression became an inescapable part of urban life in the late nineteenth century, and packages and labels were the way in which it entered the intimate realm. Packages and their labels lend themselves to formal abstraction; often they are abstract designs. Yet they lend an element of reality to a scene, a glimpse of things around the house, a hint of the time of day that is being evoked.

Most artists' depictions of packages, from Picasso and Braque to Warhol and Johns and beyond, are part of the tradition of the still life. By showing one moment in the existence of perishable things — a fleeting light, some pears, a fish — still-life paintings have traditionally commented on mortality. Packages, which promise to preserve the perishable, offer a kind of debased immortality. War-

191

hol's Campbell's soup cans help call attention to a kind of visual expression that is so much a part of modern life that it is scarcely noticed. But they also hint at the banality of what can be preserved. They are visions of a kind of prosperity that is at once comforting and hardly worthwhile. While Warhol's soup cans are full of condensed promise, Johns's coffee cans and ale cans are empty, emblems of mortality in a consumer culture.

More recently, some artists have begun to use the packages themselves as art materials. In a 1991 installation, the artist Karen Kilimnik constructed a landscape of pastel plastic bottles of beauty preparations, which was meant as a critique of how women are induced to fashion themselves. Still, one looks at the packages in the installation and sees that they are intensely communicative and full of a kind of skillfulness that was once expected of fine art.

In the late nineteenth century, there was a widespread hope that art could humanize industry. A few well-known artists did design labels at that time, and some products appropriated and commercialized well-known artworks. Rembrandt has entered countless American households via the Dutch Masters cigar box, but whether that has made smokers and their families more or less sensitive to art is difficult to say. The geometric abstractions of Piet Mondrian inspired the look of L'Oréal hair grooming products. Picasso and other modern masters designed labels for some very expensive French wine.

But for the last century, the beautiful package has been a product of popular culture, not art. As Max Beerbohm wrote in 1901: "If the ladies on the chocolate boxes were exactly incarnated, their beauty would conquer the world: yet no discreet patron of the arts collects chocolate boxes."

Package design is not fine art, as we define it today, and it does not pretend to be. And yet there is undeniably art in it that connects with much older traditions. Over the last two centuries, the quest for individual self-expression has dominated our thinking about art. Yet much of the great art of the past sought to be a collective expression or was used as an instrument of propaganda. Moreover, the creators of religious art have often had to deal with doctrinal issues at least as complex as corporate marketing philosophies, and religious bureaucracies as self-protective and politicized as a modern corporation, and have nevertheless produced transcendent works. It's even possible that some of the Renaissance masters did not have any stronger belief in St. Jerome than contemporary designers have in the margarine or fabric softener whose vessels they create.

Some commentators have argued that package design is a kind of folk art, anonymous and universal, responding to an aesthetic that is unspoken but widely shared and less directly influenced by fashion than is clothing or interior decoration. A century ago, when the owners of businesses saw their packages as their personal signatures, that may have been true. In an age of corporate bureaucracy, market researchers, and psychological consultants, there are no longer any naive designs.

No artistic or design endeavor is more compromised — or to put it another way, more disciplined — than is package design. Package designers must say what the package has to say, and just as important not say what it shouldn't say, to an audience that grants most packages only an instant of its attention. One study found that although shoppers simply do not see a lot of the packages in a supermarket, they still are aware of about eleven thousand different packages during the 1,800 seconds they spend walking the aisles. That is about one-sixth of a second per package. As one researcher expressed it, the designer has about as long as a flash of lightning to get the point across. In this very short time, there is little room for self-expression, but there is enormous need for creativity. It's possible to analyze many of the things that make a package effective. But it requires artfulness to put those elements together so that, in the blink of an eye, the package is not merely visible, but more important it means something.

Packaging belongs among those art forms — sonnets, haiku, mime, and fresco painting, for instance — that gain much of their power from their constraints. Each package has little time and little space but enormous competition from other packages. It has all the other practical obligations of protecting contents and easing shipping and handling. And within all these limitations it must communicate a complex message. As we shall see in the next chapter, packages speak strongly to the intellect and even more powerfully to the emotions. Everything humans have learned about visual expression is distilled on the faces of products.

In the packaging industry, art and aesthetics are frequently equated with self-indulgence. Yet most package designers have an art school background, often in graphic design. As we shall see in the next chapter, psychologists and market researchers have identified elements in packaging that are communicative and have strong associations. Such analysis can be helpful, but it can only go so far in the actual design. This is, though severely circumscribed, a creative act, carried out by artistic means.

Yet there is no critical tradition in package design as there is in related fields, such as product design or graphic design. Books on package design tend to consist almost entirely of carefully composed and lighted studio photographs of packages, showing them in a way in which they will never be experienced. The editor of one recent book on the subject wrote that her sole criterion for inclusion was that it had to be "a great package." Like most others who compile such books, and those who give awards for package design, she never explained what makes a package great. The tacit message of all these celebrations of package design is that a great package is self-evident. You'll know it when you see it.

This is, of course, a very permissive standard, and even it begs the question. There is plenty of evidence to suggest that the greatest packages are those that we don't really see but, rather, accept without question. Premium chocolate bars seek to distinguish themselves and command higher prices by using gold or embossing on their labels, and some of these designs have won awards. But few would disagree that the greatest package in the category remains the simple brown-and-white Hershey's bar wrapper, at once visible, communicative, and thoroughly identified with the product. Indeed, because the same graphics are stamped on the bar as are printed on the label, the package and the product are viewed almost as a single entity.

And even if you grant that the process of award giving or selecting for a book requires that the viewing be slowed down, there is still the question of how to look at a package. Do you do it in a photograph, where we might really be honoring the photographer's art rather than the package designer's? Do we look at the package in a comfortable, well-lighted room, seated at a cherry table? Any package designer will tell you that the design that looks best at the conference table often loses its power when it is placed on the shelf of a store or in the home environment where it will be used, while the package that looks unimpressive in isolation can become a star in the real world of the supermarket, pharmacy, or kitchen. Some packages look a lot better in quantity than they do singly, while others gain power in isolation. Which package is great for the product depends largely on how and where it is sold, and how visible it will be once it is taken home.

Package designers have traditionally used very long checklists of as many as three hundred specific and often difficult questions to help their clients define their goals and detailed, concrete standards for a successful package. To some extent, this is done because, in most companies, decisions are not made by specialized packaging executives, but by brand managers and others for whom package design is not part of their training or major interest. The checklists are also useful for the designers themselves, to remind them of how complex the design process is and make sure they haven't failed to address something that will haunt them later.

In nearly every case, the process includes a close analysis of competing products and forces decisions about whether the new product should resemble them or stake out a new design departure. Such a survey implies a careful look at colors to determine which ones are associated with specific competitors, and which with the product category. It also requires very clear ideas about what advantages the product has, who will buy it, and how often.

Sometimes even the question of whether the product will come in paperboard or plastic, in a bottle, or in a pouch is open to consideration. Such decisions obviously affect the graphics that will be used, the ability to print particular colors and types of images on the package. Seemingly nondesign issues having to do with materials, permeability, protection, expected shelf life, and even the expected cleanliness standards of the retail environment end up having a very strong influence over how the products look. The type of container chosen and the technology used to fill it have an obvious impact on costs, which require that detailed budgeting be a prerequisite for package design. Legal requirements for ingredients, nutrients, hazard warnings, and the like have to be considered at quite an early stage.

Retailers are the first constituency the product must satisfy, so questions must be answered about how the package will be unpacked, priced, stacked, displayed, and whether the product will be seen in large multiples or only a few on the shelf. If the product is to be bought primarily on impulse, you might design a package that hangs from a hook near the cash register, but if it is expected to be part of a shopper's planned purchases, it must be designed to overcome the less prominent display it will probably get in the store. What department of the store the product is aiming for and the typical lighting for that section also come into play. Often both market research and package design research are conducted right in the retail environment where the package is expected to be sold.

The household environment is equally important. How will the product be used in the home? How will it be dispensed? In what room will it be used, and to what extent should it respond to home-decorating trends? How often will the package be opened or closed? What sort of fears and desires should the package assuage or spark?

The issues listed in the previous paragraphs are only some highlights based on a few such lists. It's obvious that the goals and requirements that emerge from such lists have little to do with what is conventionally defined as "artistic expression." Yet such questions do underlie decisions about colors, graphics, shapes, and textures that can, in fact, be highly communicative, even artistic.

"A good package," said Richard Gerstman, of Gerstman + Meyers, one of the leading American package design firms, "has a clear, new idea." This idea can be quite simple — for example, a photograph of ice cream that emphasizes its texture and is so vivid shoppers almost want to lick the carton. "What you see on the package shouldn't be promotion, but information. The elements of the package should have a logical priority and no element of the package should appear to compete with any other." Gerstman adds,

"If you design a package that everyone else copies, that's an indication that it's a good package."

The major reason for the lack of any framework for criticizing a package is that package design truly has a bottom line. Unless a package design can get the product noticed and accepted, and thus move the product off the shelves consistently, it is a failure. Many packages that photographed beautifully, and were very pleasant to handle and use, still flunked the basic test of getting purchased again and again.

No single standard can be maintained in a marketplace of diverse, ever-changing tastes. Differences spring from ethnicity, location, age, education, and countless other factors. Some of these differences have practical origins. New Yorkers, in a constant battle against cockroaches, demand cookie boxes that can be closed, while other Americans feel comfortable with bags. Some differences grow from convictions — environmentalism, for instance — that relate to education and income but in extremely complicated ways. Youth culture, which has an impact on many products, is notoriously volatile. Few products can prosper appealing only to a single niche. Like political candidates, they must build coalitions.

Sometimes those selling the product seek to place it in a market niche to which it really does not belong. A package design that raises unrealistically high expectations can alienate those who expect a premium product, while intimidating buyers in the segment of the market to which the product properly belongs. A recent Nabisco repackaging campaign for cookies aimed at adults spurred a backlash when consumers discovered that the cookies did not deliver the experience the package promised. A package can be better than the product it contains, but not for long. It might photograph as a great package, but it will not function as one.

Nearly all packages that win awards or are reproduced in books aim for a premium market, which reflects both the larger budget available for the packaging of products aimed at upscale consumers and the class biases of the judges and editors. Indeed, an attention-getting package is something that defines many premium products. Most products, however, are not upscale, and they demand packages that people feel comfortable making part of their daily lives. Consumers make such judgments not as connoisseurs or as curators, but rather as people who would rather not be bothered.

Sometimes an attractive package can embody the wrong marketing premise for its time. During the 1960s, Johnson & Johnson introduced Micrin, a blue mouthwash in a handsome bottle that was a graspable, easy-to-handle version of an old-fashioned apothecary jar. It conveyed the message that bad breath is a medical problem that must be banished in a medicinal way. It was a variation of the Listerine approach, with more appealing packaging but without the weight of tradition behind it. At the same time, Procter & Gamble introduced Scope, a green mouthwash in a flattened conical bottle. Its color and its almost playful shape suggested that bad breath is part of life, and Scope helps. Both were backed by major corporations and huge marketing campaigns, but the times were more green and permissive than blue and inhibited. Micrin is remembered by some as an unusual and handsome package, but the product is dead. A modified version of Scope is still on the shelf and, more important, in millions of homes.

The packages that dominate awards and picture books are those whose marketing premise requires that they be admired as packages. Cosmetics, for example, are products that deal in dreams and illusions. Packaging plays an extremely important role in evoking a mood for the product, the context in which it is to be used. The packaging is often featured in advertising, so it is designed to be photogenic. The packages range from highly sculptural, jewelry-like invitations for self-indulgence to the straightforward, environmentally conscious refillable containers sold by the Body Shop chain. But whether the image sought is simple, natural girl or exotic courtesan, the container counts for a lot. After all, cream preparations intended for teenagers and grandmothers look much the same and contain many of the same ingredients. Out of their jars, cosmetics have an appearance somewhere between neutral and disgusting. But if you know that what is contained in a jar is based on an antique from the personal collection of Ralph Lauren, you might think the product is wonderful — or awful.

Attention-getting packages play a big role for products that are costly but whose benefits or distinctions are not readily apparent. Premium vodkas offer the most obvious example. Not since Uneeda biscuit has an advertising campaign been so focused on a package as that of Absolut vodka. The bottle has a very distinctive profile, and it was a pioneer in the use of transparent labels to allow shoppers to see the product within. The advertising seeks only to establish that silhouette as a prestige item, and it doesn't mention the contents at all. Consumers are encouraged to buy the bottle, not the vodka, and they do. The whole vodka aisle is a bottle beauty contest. Finlandia's highly sculptural bottle, designed in the same style as some of the renowned Ittala crystal, embodies the iciness and cold clarity associated with vodka and is sensual at the same time. Newer brands like Icy follow Absolut's lead in establishing a distinctive profile. Fris, a recent introduction, has a cylindrical bottle, whose top is sliced at a diagonal. This is a new look in liquor bottles, a belated response to the diagonally sliced top of New York's Citicorp Center.

This war of the beautiful bottles results both from vodka's lack of distinctive flavor and from its longtime associations with modernity. Liquors with heavier tastes, such as whiskey and brandy, are associated with tradition, and even newly introduced brands will offer packages designed to make them look as if they have been around for years. Bourbon is particularly prone to a folkish, neo-primitive approach; the most expensive brands strive to appear the most naive. Just as much care and expertise go into such a package, but the result is less attention getting. Among liquors, gin can occasionally be modern, but vodka alone has the license to be avant-garde.

Water, an even more characterless product than vodka, has been getting similar attention. The distinctive green Perrier bottle established the category, but there has been a proliferation of packages and feelings. The slightly clunky graphics of Poland Spring are intended to convey a down-home naturalness. The unadorned, beautiful blue bottle of Ty Nant gives a sense of old-world sophistication and, unlikely as it seems, craftsmanship. And the furniture and product designer Philippe Starck in the Glacier bottle gave form to what is probably the ultimate designer water. A pure, clear form, it has a distinctive, vaguely Batmanish black, two-eared closure, with one of Starck's trademark dangerous-looking pointed forms visible on the inside.

Motor oil seems, at first, to be an unlikely product for sensual, cutting-edge packaging. But because motor oil is a product whose effect the consumer doesn't directly experience, at least immediately, it is a good candidate for expressive packaging. Moreover, motor oil has recently ceased to be purchased primarily at service stations, where it was poured in by an attendant or a service technician, and has instead become available at general merchandise stores and supermarkets. Thus, it had to be taken out of the unwieldy tinplate can and put into something else. Nearly all brands of motor oil have been repackaged in highly sculptural, asymmetrical plastic containers with tops designed to minimize spilling and the phenomenon known as glugging. The most frequent look is hypermasculine high tech, with the plastic given a metallic sheen. In several cases, the small container is shaped to fit the hand when pouring, with a dimpled surface to prevent it from slipping. Castrol's Neutron X was given a large cap that resembles an oversize bolt head. The larger sizes of BP have ridges that assist in pouring, but which also resemble the surface treatment of automobile engines themselves.

Recently, there has been a tendency to tone down the strident, high-tech imagery of these packages, perhaps because women are important purchasers of motor oil and are intimidated or repelled by the Robocop machismo of these containers. Products that are new to the shelf often start out with very expressive packages and then calm noticeably as the product category wins consumer acceptance. We have seen how that happened with liquid detergent. Motor oil, while not a new product, is still relatively new to self-service retail environments, and it seems to be undergoing a similar transition.

*　　*　　*

Sometimes, the kind of attractive, noticeable package that wins awards also establishes itself in the supermarket mainstream. One notable example during the 1980s was the Classico line of pasta sauces, a new product line whose packages were designed by the Duffy Design Group. The labels contrast with others in the category by using illustrations of landscapes identified with the region of Italy for which each sauce is named. The jars themselves resemble old-fashioned canning jars, which suggest that although the product is not homemade, it implicitly competes with the sauce-making grandma most of us never had. The jars are smaller than those of competing sauces, to suggest preciousness, and the prices a little bit higher. The jars also encourage reuse for storage and thus reduce the guilt induced by throwing things away. As with many new products, there was very little initial advertising, so the package had to draw shoppers' attention and make the sale. The package and the sauce together have convinced shoppers of their value, because Classico has maintained its place on the shelf and its higher price in a category that has been subject to heavy price-cutting and substantial inroads by retailers' private-label brands.

Commercial success makes it easier to conclude that Classico, Absolut, and Hershey have great packages. What's tougher to judge is a product like Scope, which is a long-running success but whose package is not visually distinctive. As was suggested above, its ordinariness played a major role in its success, not only against Micrin but against the dominant product in its market, the aggressively medicinal Listerine. Most of life is ordinary. Nobody wants a houseful of products that are as self-dramatizing as operatic divas. Only rarely, when some special reassurance is required, do people need beautiful packages. Most often, they want packages they can live with.

In earlier chapters, the Wrigley's Spearmint gum wrapper was cited as an outstanding package. Its bold green arrow is the kind of simple, brilliant graphic idea that designers today are too sophisticated to propose. But it achieves quite a lot. Obviously it calls attention to the package itself, making it stand out among the many equally small products with which it is most often displayed. The use of the large-scale element on the very small package gives it a sense

201

of pent-up energy, which is enhanced by the directionality and dynamism of the arrow itself. The arrow is also a kind of abstracted spear, a reminder of the spearmint flavor. The green of the arrow communicates mintiness. Spearmint gets its name, of course, from its pointed leaves. But by abstracting this image, the designer of the package was able to communicate forcefully, though not stridently, all the coolness and energy that a buyer hopes to get from this little pack of gum. In recent years, Wrigley's, like Coca-Cola, has increasingly relied on packages that offer more of the product than their traditional pack, but which are less powerful packages.

It's important to note that while the Wrigley's packet, the Marlboro box, and the Hershey's wrapper are graphically quite simple, this simplicity is not an end in itself. Sometimes, with packaging, less is more, but often it's not. These three products are all small and frequently displayed with many other similar products. A simpler package stands out. They are also, to some extent, purchased repeatedly, as a matter of habit, so they needn't sell too hard.

What makes the Wrigley's package outstanding is not its simplicity but its communicativeness. The green arrow is the perfect symbol for a product that offers the primal comfort of chewing with a stimulating taste. The package doesn't so much sell the product as indicate how to enjoy it. The more ornate Classico jar does much the same thing. It encourages people to taste the sauce not as the common commodity it has become but rather as an expression of a place, with its own culture and distinctive ingredients. It makes people feel that they are not simply opening a jar of sauce but doing something just a little more special. The communicative qualities of very good packages are not strident sales pitches. Rather, they establish a relationship with those who use the product. And when we say that packages are important cultural expressions, it is not to suggest that the culture is dominated by hucksterism, but rather that packaging provides a way in which people define and understand themselves.

That doesn't mean that packages are profound, that they replace religion or philosophy, literature or art, as explorations of values, meaning, or purpose. They are collective efforts, deeply compromised. Any temptation to view them as artworks, products of a personal vision, can be overcome simply by looking through a design firm's archive at the designs that were not produced. Sometimes there have been dozens of designs for a single product, exploring every way in which it can be expressed. For automatic dishwashing powder, for example, you can have an illustration on the box, or try to stand out from the competitors with a photograph. The illustration can be abstract or realistic. You can show a gleaming glass, like everyone else, but the glass can be modern or antique crystal. Or you can find some kind of a distinctive glass or dish and let it sparkle. You can settle for simply showing the dish, or you can also show the satisfied expression of the woman (men still aren't an option) who washed the dish. You can choose colors that give the box a clean, scientific look, or you can give it some kitchen warmth. For some clients, design firms can and do go on and on, providing different designs of almost equal visual polish, but each of which says something subtly different about the product.

Those who produced the designs may have their favorites, and if they are any good, they can probably pick a few that will stand out on the shelf, fit into the home, and meet the client's marketing objectives. But while the designers have a sense both of professionalism and craftsmanship, they are under no illusion that the way in which they represent dishwashing powder will be a significant artistic expression or a lasting statement about late-twentieth-century life. Sometimes, though, that's exactly what happens.

8 *Seeing and believing*

*T*hou still unravish'd bride of quietness," wrote Keats, in the first line of his *Ode on a Grecian Urn*, addressing a protopackage. He looks at the ancient Greek urn, which represents something distant from himself and his own experience, and sees a scene of ancient ritual and sacrifice, probably inspired by the frieze of the Parthenon in Athens. But what he is really looking for is himself. The urn's role is to excite his own sensibility.

Although he goes at it with greater intensity and higher expectations than a child reading a cereal box at the breakfast table, Keats expresses a similar faith that a container can tell more than merely what's inside. Keats finds transcendence in what we might now characterize as a scene of blood sacrifice and implied rape. The child might be contemplating a box of Count Chocula cereal, eating as he gazes, engaging in a sugar-coated communion that allows him to participate in this month's pop-culture myth.

People participate in packages, and what they find in them is often far more than what they see. Psychologists know that we hear their unheard music, project ourselves into the scenes they depict, and are energized by the dynamism of what we find on them. Like

Keats's urn, packages are repositories of feelings and values — even if we do not look at them quite as closely as Keats studied the ancient vessel.

Keats's ode isn't meant to be about packages, but it does follow an emotional trajectory that package researchers are familiar with. The poem can be said to have three distinct movements. The first few lines constitute a near-ecstatic cry of discovery. The urn is an unravished bride, a foster child of stillness, a sylvan historian more expressive than the poet. In this opening exclamation over the urn, Keats is jumping to conclusions, and he engages the reader with his enthusiasm, though the reason seems unclear. During the second movement, which constitutes most of the ode, the poet describes the varied scenes on the urn and reflects on each of its aspects. The third movement, which comes at the end of the last stanza, contains Keats's ultimate response to the vessel. It ends with the famous lines " 'Beauty is truth, truth beauty,' — that is all / Ye know on earth, and all ye need to know.''

To a literal-minded reader, it is not altogether clear that this conclusion is justified by what has been described earlier. In fact, the conclusion is far more of a piece with the cry of discovery in the first few lines. The poem is structured as if it were a rational argument, but it is really a flow of emotion. What is depicted on the urn is not so much the reason for the poet's response as its occasion.

This same cycle of emotion-laden discovery, examination, and ultimate judgment — a judgment that is more closely tied to the emotion than to the examination — has been found to characterize the shopping experience. Since the 1930s, package design has been subjected to a substantial amount of psychological research, which has tended toward the conclusion that shopping is an irrational process and that packaging is effective primarily insofar as it appeals to the subconscious. The retail package is Keats's unravished bride — virginal, immaculate, but ready to be possessed, unwrapped, used by the consumer. The first package the shopper examines is almost always the one that is bought. The instantaneous emotional reaction carries the greatest weight.

205

*　　*　　*

During the same period, and especially since the rise of consumerism during the late 1960s, there has been a countermovement to require that packaging educate the consumer. In most countries multiple government agencies regulate claims of healthfulness and efficacy of products, and many require the inclusion of nutritional information and warnings of possible allergic reactions on the packages.

While some package designers might accept Keats's ultimate identification of beauty and truth, those who advocate greater truth in packaging certainly do not. Emotion speeds judgment, though some feel it does so by accelerating thought, while others feel that it short-circuits intelligence. In any event, today's packages embody conflicts between reason and emotion, government and business, words and colors, form and content.

People are affected by packaging in/ways that they do not consciously understand. On that both packagers and consumer advocates agree. The controversy is whether to see this as a key technology of our consumer society or to try to work against the power of packaging and slow shoppers down. Packages are everywhere, and by their nature they contain information as well as products. Some of this information consists of words and numbers, directed to the rational mind, while other facets, consisting of shapes, colors, and graphic expressions, bypass the rational and appeal directly to consumers' emotions. Consumer advocates focus on regulations that enable people to make rational decisions. Package researchers and designers say they see labeling regulations as inevitable and, in most cases, desirable. Such information helps product manufacturers to show that they are responsible and on the consumer's side. But the researchers have plenty of evidence that reasoned analysis plays only a small role in most shoppers' decisions.

Packages currently available in American stores are typically the work of designers, engineers, and one or more government agencies including the Department of Agriculture, the Consumer Product Safety Commission, the Environmental Protection Agency, the Federal Trade Commission, the Department of Health and Human Services, the Bureau of Alcohol, Tobacco, and Firearms, and the Customs Service. During the 1980s, despite political leaders who were rhetorically hostile to regulation, the amount of information on packaging substantially increased, and at least some of the enforcement authorities were aggressive in pursuing false or exaggerated claims.

The recent, massive relabeling effort was one of the most dramatic successes by consumer advocates in a long history of attempts to use the communicative power of packaging to impart useful information. The idea is to use packages to provide objective grounds for judgment. It is an extension of the principles of the Pure Food and Drug Act of 1906 and of the failed 1933 attempt to impose uniform grade labeling on all products. Food manufacturers fought grading, but they did not lobby strenuously against the new regulations. In 1992 they came to a fairly rapid agreement with the federal agencies involved in nutrition labeling, in part because they were eager to resolve the issue before a Democratic administration took office. But there is little doubt that industry's docility stemmed in part from its understanding of how package design affects people.

The principal impact of the new regulations for the great majority of packages was the inclusion of a standardized box headed "Nutrition Facts" in blocky boldface type. The box contains such information as total calories per serving, calories from fat, and the amounts of fat, cholesterol, sodium, carbohydrates, and protein. Below, in smaller type, it contains the percentages of daily requirements of certain vitamins and recommended intakes of these nutrients. Thus, every food package now contains a condensed guide to overall diet. Most food products had contained some of this information before. What the new regulations did was standardize such issues as serving size, make the information as legible as possible, and seek to give the shopper a basis for comparison.

The regulators' insistence on a uniform graphic form, with specifications for the width of letters as well as their height, shows sophistication. In the past, designers have tended to put all undesirable information in highly condensed, illegible type. On almost all products, this newly required nutrition information box takes up more space than nutrition information did previously. The new box is, itself, designed in a more assertive manner than before. Thus, food manufacturers were faced with a large, intentionally noticeable new element that is not unique to their packages and that, by encouraging comparison, seeks to undermine the mystique that package design tries to create.

But while getting the complex "Nutrition Facts" box onto the package was frequently a nuisance for designers, manufacturers were largely unconcerned, precisely because of the large amount of information it contained. The "Nutrition Facts" box requires analytic thinking, and even mental mathematics. Most shoppers do not even keep track of how much money they are spending, let alone tally the number of grams of saturated fat for each anticipated menu. Most purchasing decisions are made in a tiny fraction of a second, so there is little likelihood that increasing information at the point of purchase will increase the rationality of consumers' decisions.

That's the major reason the Food and Drug Administration has resisted an initiative by the American Heart Association to put its logo on products that have met its standards as healthy foods. The logo would be easy to read, and officials have said they fear its presence would cause shoppers to ignore the nutrition facts and make quick decisions based on the endorsement, without regard for how they will use the product as part of their diet. Regulators hope to slow shoppers down, not speed them up.

Most of the grocery manufacturers' complaints about the new labeling regulations concerned the new standard definitions that are being applied to the use of words like "light," "lite," "low fat," and "reduced." These go on the front of the package and can be read in the tiny split second that the shopper devotes to the package. In this way, it is more like grading products A, B, or C. Some products had to be reformulated to be able to continue to use their name, while for others, there was little choice but to change the name or leave the marketplace.

Package designers generally expect that the regulations' chief impact will be on products that are sold on a nutritional basis. While shoppers generally grab packages automatically, they sometimes scrutinize products carefully to make comparisons, and the labels will enable shoppers to weigh, for example, Healthy Choice against Lean Cuisine. And the impact of nutrition information might well be greater at home than in the store. Research has shown that people really do read cereal boxes. If a cereal that is being sold as healthful and nutritious turns out to contain a lot of fat, it will eventually be noticed.

Although the expectation that the new nutrition label will be studied at the store is unrealistic for most packages most of the time, this first encounter on the shelf is only the beginning of shoppers' relationship with the package. At home, there is more time to read the material, and packages are not arranged in the cupboard with their selling face forward. Manufacturers use the sides and backs of packages to include complex information, such as recipes, which are intended to accelerate use of the product and induce repeat purchases. Thus, the complex information of the nutrition box is not necessarily wasted, especially if it helps consumers make decisions about what to purchase next time.

"If one wants to act rationally, one must, at all costs, find a reason which makes the irrational seem rational," wrote Ernest Dichter, a package designer and author. Some of what appears on the package — especially the words — are there to help reassure consumers that their impulsive choice was also a sensible one.

Thus, nutritional information on the package won't be emotionally neutral. Anything that's put on a package — even a bunch of scientific names and numbers — can trigger feelings as well as thoughts.

"One peculiar impact of the new requirements is that products that have essentially no nutritional benefits at all will have a label that says 'Nutrition Facts,'" said John Lister, president of Lister-Butler, a leading design firm. He argued that the mere association of junk food with the word "nutrition" might give it more respectability and ease people's guilt in purchasing the products. He added, though, that it would eliminate past abuses, such as cracker packages that showed crackers and soup together as a "serving suggestion." "The nutritional chart looked terrific," he said, "but all the crackers contributed was flour and sodium." The rest of the nutrition came from the soup that was pictured on the package, but wasn't contained inside.

Such a tactic wasn't, strictly speaking, dishonest. A careful reader could understand that the information referred to the total meal, not simply to the crackers. The problem is that most people don't read a package as if it were a legal contract. Usually, they don't read the package at all.

"I once attended a meeting at 7Up where the designer of a proposed new can had omitted the word 'Up,' " said Davis Masten, president of Cheskin + Masten, a top psychology and marketing consulting firm. "The meeting went on for over an hour before anyone noticed what was missing." If the executives of a company fail to notice that two-thirds of their brand's name is missing from the can, it's understandable that people rushing through a super-market might miss some of the finer points of the thousands of packages they pass by, or even the few dozen they put in their cart. People don't read or even look closely at most packages. They don't analyze, but they see more than they know.

"It should be understood that people have built-in defense mechanisms against words," Bonnie Law of the Color Research In-stitute of Chicago wrote in 1981. "People are not defensive in re-lation to forms (images, symbols) or colors because people are not aware that they are affected by forms or colors. To consumers, the importance of a package is to serve as a container for the product and the purpose of the label is to identify the product." Indeed, in many cases, the letterforms of the words themselves have been shown to have more meaning and more impact for consumers than what the words actually say. Reading the words on the package is one of the last things people do.

The pioneer in studying people's emotional response to packages was the marketing psychologist Louis Cheskin, who began his research in the 1930s. He was long associated with the Color Research Institute and he was later immortalized by Vance Packard as the most articulate and engaging of his hidden persuaders. His seminal experiment on packaging involved placing an identical product in two different packages, one identified with circles on the outside, the other with triangles. He didn't ask his subjects to say

Vodka and water are characterless products that depend on their packages to assert premium status. Absolut vodka (A) has successfully made its package an icon. Stolichnaya (B), a throwback, still has a factory on its label. Finlandia (C) is crystalline and cool, while Frïs (D), a new entry, goes for hard-edged elegance. Waters include French classics Perrier (E) and Evian (F), along with Ty Nant from Wales in its sensuous glass bottle (G) and Glacéau (H), packaged as an accessory to an active life.

18

A

B

C

Recorded music began with the Edison cylinder, packed in a cylinder of its own (A). The long-playing album provided a large surface for memorable graphics (B), while the compact disk long-box required destruction of art and materials before listening (C,D).

D

Stopette deodorant, the first product to come in a plastic squeeze bottle, was introduced just as World War II ended (A). Aerosols, developed during the war for insecticides, were quickly applied to other products, such as cheese (B), whipped cream (C), and perfume (D).

20

A

Dutch Masters turned Rembrandt's art into packaging (A). Andy Warhol turned Campbell's packaging into art (B).

Visual styles often have an effect on packaging, such as a machine-age 1930 Wesson Oil tin (C), and a mid-1980s lipstick line that evoked the skyscrapers of the period (D).

B

C

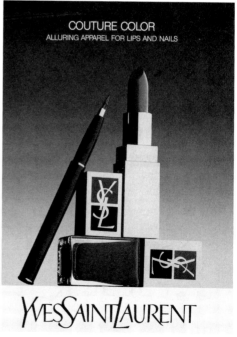

D

A

B

C

D

Packaging is a large part of the fascination of perfume (A). Houbigant (B) was a latter-day cubist package. Some French bottles, such as the Lalique-designed Air du Temps and the modernist Chanel (C), began as stylish and subsequently became timeless. The recent Minotaur (D) seeks to fuse neoprimitive styles with the classic.

22

A

B

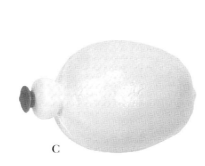

C

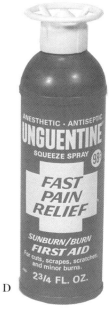

D

E

F

G

H

I

Reliquary of Saint Bylbos (1467) (A) does not hold the bishop's arm but rather embodies his episcopal gesture. Electric spinach label (B) likewise evokes power more than contents.

Sculptural packaging represents its contents, like the squeeze lemon (C); deals in metaphor, as does the burn medicine (D); reminds people of a current brand name, like the Shell lighter fluid (E); or creates a new brand identity, as did L'eggs pantyhose (F).

Aunt Jemina pancake ready-mix was an early example of a product that packaging made possible (G). Some others: direct-to-table salt shakers (H), air freshener (I), stovetop popcorn (J).

J

24

Some contemporary packaging trends: boxes and glass bottles replaced by a flattened plastic bottle with major graphic impact (A,B); concentrated refills in lightweight containers (C,D); new containers that use minimal, often recycled materials (E); new closures that make life easier for the old or disabled (F) or help everyone understand how to open a reclosable bag (G).

anything about the packages. He wanted to know which product they preferred and why. He found that 80 percent of his subjects preferred the product in the box with the circles over the one with the triangles. The reason they gave was that the box with the circles was a higher-quality product than the box with the triangles — even though the contents were identical.

"I had difficulty believing the results after the first 200," Cheskin wrote later, "but after 1,000, I had to accept that many of the consumers transferred sensations from the circles on a carton cover . . . to the contents of the container."

Cheskin then varied the test. He gave his subjects the two cartons and asked them to predict, before trying the product, which one would be better. They preferred the product with the circles by a similar margin. Then he asked them to try the product and indicate their preference. Their actual preferences differed from their predictions by less than 2 percent. Cheskin repeated the experiment for a wide variety of types of products, and he found that the look of the package has an enormous impact on how crackers taste, on how soaps are perceived to clean, on how rich and satisfying a beer is to drink. He named this phenomenon "sensation transference." It became the foundation not only of Cheskin's career as a consultant to such companies as Procter & Gamble, Standard Oil Company of Indiana, and McDonald's, but of much of the research in package design that has been done since.

"Two concepts form the basis of package design research," wrote Walter P. Stern in a textbook on the subject published in 1981. "One, consumers really do not distinguish between a product and its package — many products are packages (and many packages are products). Two, consumers relate emotionally not to the facts (the realities) of the products/packages they are involved with, but rather to their perceived realities." In other words, all package design research, no matter what methods are used, is about Cheskin's principle of sensation transference. It has little interest in what people think about the package. What's important is what the package makes people feel about the product.

Cheskin's original experiment has been repeated year after year, with countless products in a great many variations ever since he first did it. And despite increasing consumer sophistication about marketing tactics, despite defensiveness and cynicism, it still works. As Stern noted, a blind test of beers can be repeated over and over again, with a strong consensus about their taste and quality. But when the beer bottles are introduced, they change the way people taste the beer.

One of the most dramatic versions of the Cheskin experiment involved an identical underarm deodorant that was mailed out to a test group in three different packages in which the color combinations were varied. The subjects were told that the deodorants were three different formulations under consideration and were asked to state their preferences and their judgment of each of the three formulations. In such three-way tests, consumers sometimes react like Goldilocks did to the Three Bears' lifestyle. In this case, color-scheme B was the one considered "just right." Those tested praised its pleasant yet unobtrusive fragrance and its ability to stop wetness and odor for as many as twelve hours. Color-scheme C was found to have a strong aroma, but not really very much effectiveness. And color-scheme A was downright threatening. Several users developed skin rashes after using it, and three had severe enough problems to consult dermatologists.

Recently, Cheskin + Masten, the successor firm to one Cheskin founded, repeated a variation of the original experiment for the makers of Christian Brothers brandy. Four different groups of brandy drinkers were used. The first group compared brandies in a blind taste test, in which neither Christian Brothers nor E&J, a comparably priced, California-made competitor, was identified. Christian Brothers won, with the respondents saying that it had a "more upscale" taste. The second group was given the brandy identified by brand names, but without the bottle. The preference for Christian Brothers was more overwhelming, with Christian Brothers praised for its "smooth taste." The third group was served the brandy from bottles. The Christian Brothers bottle was plain and rather generic. The E&J bottle, while not the sort that typically wins design awards, had an ornate, old-world look. This group picked the E&J as the higher-quality product. The fourth group was served E&J from the Christian Brothers bottle and Christian Brothers from the

E&J bottle. They had a strong preference for the brandy in the more elaborate bottle, in this case Christian Brothers. The clear conclusion was that Christian Brothers would benefit from a new bottle. The company later improved its sales by introducing a fancier new bottle, almost as ugly as E&J's. The consultants had advised a far less restrained approach than the company finally adopted. Nevertheless, Cheskin had been vindicated once again.

This example demonstrates that there are ways in which this key experiment can still be useful more than half a century later. But the main thing it shows is that package design is important. Only indirectly and through a great deal of trial and error can this kind of experiment produce insights into what elements of a package elicit desired or counterproductive responses. Thus, Cheskin's work opened the door to a variety of analytic approaches, ranging from the physiological to the aesthetic.

Cheskin himself specialized in a kind of visual analysis, involving shape and color. Like the members of the first generation of industrial designers, with whom he sometimes worked, he presented himself as a combination of guru and magician, offering visual approaches to solve business problems. Unlike the designers, however, he had theories that could be tested by experiment. He had not reduced design to a science. But he did at least provide some scientifically based tools for evaluating it and some principles that he had extracted from his research to give designers a head start. The designs of the Oxydol and Tide boxes, discussed earlier, were based directly on Cheskin's ideas about shape and color on packages. Cheskin was particularly proud of the Tide box. In his lectures, he liked to show a slide of the Tide box that had been cut into sixteen squares that were then randomly reassembled. It was still immediately recognizable as a Tide box, one of the strongest designs for brand identity yet devised.

Many of his ideas about design had already been in common use before he began to systematize them. They involve geometry and color combinations, issues that have always been addressed by artists. His approach has a strong commonsense feeling to it.

213

What was new in his discussion was his ability to prove how potent this sort of communication can be.

The Tide box is a good example of a package that tries to reconcile two apparently contradictory qualities, power and mildness. While this conflict is peculiar to cleaning products (and some related things, such as deodorants), almost every package embodies and resolves some sort of conflict. Moreover, the resolution of opposites is precisely what can make a package compelling.

The most common source of tension in a package design is between the need to be noticed and the need to be accepted into people's homes and lives. A certain aggressiveness is useful for getting noticed, but it can affect the way people feel about the product, especially when it is removed from the competitive environment of the store and brought into presumably more peaceful domestic surroundings. It happens that the most noticeable shapes are triangles and other pointy figures involving acute angles. Cheskin's first experiment proved something important about triangles: people can see them but that doesn't mean they like them. The most noticeable color is yellow, which, for most products, also has negative connotations.

Cheskin found that a circle or an oval has the most positive association, but alone, each lacks personality. The circle or oval must somehow be inflected with some other symbolic form or identification. Thus, Tide's concentric circles are played against bold lettering, and the oval of the Amoco logo is bisected by a torch and filled with the company's name. With most packages, the rounded shape is not expressed quite so literally. But images of completeness, receptiveness, and enclosure — feminine forms — provide the underlying theme for a majority of packages. Cheskin worked with McDonald's at the time it was about to abandon arches as architectural elements of its outlets. He advised that the memory of the arches be kept in the form of the *M* in "McDonald's." His case was based, he said, on research that showed that "the arches had Freudian applications to the subconscious mind of the consumer and were great assets in marketing McDonald's food." In other words, McDonald's represents a womb, which is advantageous if you're replacing home cooking.

As you walk around the store, you'll also see many importuning, pointy, explosive masculine forms, often in bright, unavoidable yellow. But these strident shapes are less often part of the underlying package design than superimposed on it, with messages like "New and Improved," "29 Cents Off," or "Free Offer!" These, and similar elements that don't read as part of the basic package design, are known in the business as violators. On some packages, they do come and go, while other packages that depend more on impulsive purchasing hardly ever go unviolated. And when the product is brought home, it is valued for its soft shoulders, not for its sharp elbows.

If the matter was as simple as the last few paragraphs have suggested, all packages would have rounded shapes and blue labels with yellow star bursts. Fortunately, things are a great deal more complicated than that. Each kind of product has its own set of color expectations and a range of expression that is considered meaningful. Moreover, there are many ways to represent the excitement, reassurance, and satisfaction that are psychologically a part of any package.

Color is unquestionably the most potent tool for emotional expression in packaging. Studies of involuntary physical reactions — eye movement, neural activity, heart rate — show that color is the element of a package that triggers the fastest and largest response. It communicates at a level that is nonverbal and unconscious, and it is also beyond the law. Words can be regulated, and so can pictures, but color cannot.

Take, for example, the label of V-8 vegetable juice. For decades its label has shown the eight vegetables — tomato, celery, carrots, beets, parsley, lettuce, watercress, and spinach — from which it is made. Tomatoes predominate in each generation of this image, because the law requires that the ingredients be shown in roughly the proportions in which they are found in the product. Finding the watercress in the picture is, by contrast, a bit of a puzzle, which is also an accurate reflection of the juice. The general arrangement, a horizontal array of tomatoes, defined by greenery and punctuated by vertical celery and carrots, has stayed more or less the same whether the label was an illustration or a photograph, as it is at this writing.

But what you might not notice, but will probably feel, is the intensity of the colors of the vegetables. They are not printed with the standard four-color process commonly used to print color in magazines and books, but with five colors. This permits some of the vegetables to take on a purer hue than is available with conventional printing. The colors in the printed label are not necessarily stronger than would be found in reality — though these are admittedly vegetable-beauty-contest winners. But they are surely more vivid than they would be in conventional four-color reproduction, which has become a sort of standard for reality. Thus, there is something mysteriously compelling in this representation, an analogue for the energizing healthy refreshment the product claims to offer. Even though the label is in large part a collection of required and regulated words and images, color breaks through these restrictions and makes a very powerful statement of its own on the shelf. (Obviously, this cannot be duplicated in a magazine advertisement, or in this book, for that matter.) There is nothing dishonest or ethically questionable about this tactic. Color is the fundamental language of packaging, and this label tries to use it powerfully.

Color is a subject that has generated an enormous amount of portentous and empty discourse. Part of the problem is that color seems to be wholly dissociated in the mind from verbal skills, and vocabularies are inadequate. Shadings can make an enormous difference in feeling and association, which can be experienced, but not easily described. And when colors are seen together, they affect each other and induce reactions different from those that would be sparked by any of the colors singly.

People experience color in packaging at three different levels: the physiological, the cultural, and the associational. The first is universal and involuntary. The second arises from visual conventions that have grown up in various societies over long periods of time. The third relates to color expectations on packages that have become associated with a particular product category through the marketing process. This includes, for example, the yellow found on margarine boxes, echoing the coloring found in the product itself, and the yellow found on Kodak boxes, the result of close to a century of salesmanship. In practice, though, it's difficult to separate these three kinds of color perceptions.

The physiological seems the most straightforward. We know that walking into a bright red room speeds the pulse and that green can slow it down. Blue is widely perceived as calming, and there is evidence to suggest that it is. Yellow is very successful at drawing the eye, though the experience is not always perceived as pleasant. It is difficult to be sure of the universality of such reactions, however. There is a natural bias in marketing tests toward people who can afford to buy the products, something that itself counts out most of the world. And because the testing equipment is often cumbersome and uncomfortable, data are usually based on relatively few subjects.

Involuntary reaction tests, which still require human participation, measure the opening of the pupil, respiration rates, perspiration, changes in the tension of the larynx, brain waves, and other electrochemical activity in the nervous system. Eye-tracking tests, which were first used as part of the training of British antiaircraft gunners during World War II, are the most common and accepted of the involuntary tests because they can be used to analyze discrete elements of designs. Eye tracking, however, studies not the first decisive glance at a package, but rather the second and third looks, which most people don't give most packages.

The weak link in involuntary-reaction research is that the connection between the body's reactions and actual human behavior is elusive. Nobody questions the reality that people have visceral responses to colors and shapes. But just how a particular brain-wave pattern corresponds to the purchase of a pound of bacon or a jar of moisturizing cream is not so well understood. There is not a clear connection between the ability to make a person sweat and the ability to sell something. Involuntary reactions, presumably shaped by hundreds of thousands of years of evolution, are more often triggered by negative events than by positive ones.

It is easy and tempting to speculate on the origins of some well-known color responses. You can see the heat and emotion of red as a pool of blood, reassuring green as a verdant place with plenty of food, calming blue as the sky in good weather, attractive but irritating yellow as the sun. What's more likely, though, is that early humans learned to observe a great many subtleties in the colors of things around them and judged colors in terms of particular contexts. There is no one blue or red or green, but hundreds of them, all of which can have a meaning. For modern man, far less in contact with the natural world, this body of knowledge has been incorporated into culture.

Different cultures have disparate attitudes toward colors. One familiar example is that black is the color of death in Western societies, while death is colored white in many Asian countries. But even within a particular cultural context, the situation is more complicated than that. In his 1969 book, *Color Sells Your Package*, Jean-Paul Favre, a marketer for Nestlé, the Swiss-based global food company, demonstrated the contradictory ways in which people see colors:

> Black is dark and compact, being a symbol of despair and death. Its character is impenetrable. It is a void without any possibilities, an eternal silence with no future, without even any hope of a future. It is the color with the least resonance, the expression of a rigid unity without any peculiarity of its own. Black confers an impression of distinction, nobility and elegance, especially when it is shiny.

Favre is correct that in most of Europe and the Americas, black is both lugubrious and luxurious. He makes no attempt to reconcile these different connotations of a single color. You might expect a wide range of associations with colors in which there are many possible shadings, but black is more or less invariable. Nevertheless, from inside the culture, it is fairly easy to distinguish the noble, elegant black from the funereal black. It's not easy to explain exactly how to tell the difference, but you can.

The upscale image of black is, in fact, very likely the ultimate case of an often noted phenomenon of color preference. In most Western societies, bright, pure colors are loved by children and by the poor. Wealth and education bring with them a taste for subtler, grayer shades, as if greater discernment must be accompanied by sensory deprivation. The rider in the shiny black limousine inhabits a different world than most people. The black case of a high-tech sound system promises competence beyond human understanding, possibly even better than the ear can hear. The black package of cookies implies a product that's not sweet or childish, created for a palette that has moved from pure sensation to fine discrimination.

There is no other country where packages that are pale and tinged with gray are so prevalent as they are in Japan, a country whose national identity is tied to the appreciation of extremely subtle distinctions. The Nippon Institute of Color and Design Research charts color choices on a three-dimensional matrix of cold and hot, soft and hard, and gray and pure, and a very interesting pattern emerges. Japanese designers use soft, gray shades for products like tea and seaweed that are viewed as having a distinctively Japanese character, while they do not hesitate to use brighter, purer hues for products such as coffee and soft drinks, which, while widely used, are nevertheless perceived as foreign. Carefully expressed humility is a very important goal both of traditional Japanese wrapping and much of the country's contemporary packaging. By contrast, the Japanese expect foreigners to be brash, and their designers and manufacturers are willing to be even more aggressive than their Western counterparts when their task is to promote a product that is perceived as alien.

Clear, eye-catching yellow evokes from the Japanese the same adjective — "cheap" — that it does among Americans and Europeans. Its sheer luminosity and vibrancy appear to deprive it of character. People seem to vaguely resent how powerfully it can win their attention. Moreover, like the sun itself, it is perceived as borderless. A red circle has presence, but a yellow circle does not, unless it has a substantial color, such as blue, to define it. Yet it is one of the most used packaging colors, and not only for cents-off promotions.

219

Its association with lemons, sunshine, and egg yolks helps take people's minds off the cowardice and lack of moral fiber that is so often associated with the color.

One of the world's most famous commercial images, McDonald's arches, illustrate one of the perils of discussing the emotional dimensions of color. By any objective standard, these are bright yellow in color, as clear and optically unavoidable as they can be, which is one of the things that makes them effective. But nobody thinks of these arches as yellow. They have been transformed by a bit of verbal alchemy into golden arches, and thus something that had a potentially cheap connotation gains a precious overtone. Usually, gold is a slightly browner color than yellow, but this shading is missing from McDonald's. Decades of assertion that the arches are golden has made them so in consumers' minds. And with this comes an association of purity and changeless high quality. Similarly, Kodak, with its gold line of amateur films, has begun to try to turn its yellow, which is somewhat more golden than McDonald's, into gold. (Gold itself, of course, gets some of its preciousness from its dazzling color, which is impossible to ignore. That is, it benefits from being yellow.)

It is interesting that, except for its concentration on the gray scale, the Nippon Institute of Color and Design Research matrix would serve equally well in Western countries. Red, for example, might have diverse cultural associations, but there is widespread agreement that it's hot, while blue is cold, and green has a pleasant neutrality. The institute's basic research technique is to force subjects to apply adjectives to particular colors and combinations. And while a high percentage of responses concern class distinctions that apply primarily to Japanese society, those that have a sensual dimension — soft, dry, citrusy, solid — might also have been chosen by an American or European. In part this is because a lime, for example, is the same color anywhere, and a lime-green package can call up a sense of tart fruitiness no matter what language you speak. This is an associational use of color, which, as multinational companies have learned, can transcend culture, as long as colors that violate local taboos are avoided.

*   *   *

Thus, when people respond to color, they do so on all levels at once. A 1987 study of residents of four American cities found that about 59 percent of them believe that the color red represents danger, something you might expect from the physiological response. But when asked to think about products, 71 percent said red represents Coca-Cola. Perhaps the flutter of excitement one feels when confronted with danger is an appropriate accompaniment to a product that has traditionally contained caffeine and has been sold as a refreshing, invigorating drink. The same study indicated that red can also mean love, safety, strength, and warmth. Its association with safety, however, is less strong than its association with danger. Thus, while it's possible for a company that wants to be trusted to use red as part of its imagery, there is always some uncertainty and ambiguity. A dessert product that wants to convey warmth and love (along with sweetness and the likelihood that it's fattening) can do very nicely with red packaging. But the parent company, which wants to be trusted and raise its stock price, is likely to go with blue.

In this survey, deep blue was by far people's favorite color, with 41 percent listing it. Red was second, with 16 percent. Deep blue was associated with trust, good taste, quality, value, strength, and friendliness and was by far the most desired color for corporate imagery and financial institutions. But the adjective with which blue was most strongly identified was "cold." For some sorts of products, it needs to be warmed up, usually with red. Thus the common color scheme of red, white, and blue, which is used, for example, by many of the world's airlines, has little to do with American (or French or British) patriotism. Rather, the combination represents excitement mitigated by trustworthiness, cold competence leavened with care.

Favre's more subjective, not to say obsessive, meditations on color meanings nevertheless show something of the complexities of thinking about color. He first says that red "signifies strength, vivacity, virility, masculinity and dynamism," which is the theory behind wearing a red tie (with a blue suit) to a job interview. "It is brutal, exalting or even unnerving, imposing itself without discretion," he adds. "It also gives an impression of severity and dignity as well as of benevolence and charm." Having thoroughly confused

221

the issue, he then tries to give these apparent contradictions some system:

> All the shades of red have their own psychological character. Scarlet is severe, traditional, rich, powerful and a sign of great dignity. A medium red embodies activity, strength, movement and passionate desires. It confuses and attracts us. These shades of red are used when we want to indicate the primitive strength, warmth and efficiency of the stimulating and fortifying properties of a product. Cherry-red has a more sensual character. A still lighter red signifies strength, animation, energy, joy and triumph. Summing up, it may be said that red becomes more serious, deeper and problematic the darker it is and has a happier and more imaginative temperament as it becomes lighter.

Thus, when people see red, it's all this and Coca-Cola, too.

Obviously, shadings count. The 1987 American survey indicated, for example, that green is associated strongly with nutritious, healthy, natural food. This hardly seems surprising, but it flies in the face of long-held conventional wisdom that green is associated with moldiness and decay and thus very tricky to use with food products that are not themselves predominantly green in color. Perhaps this changing perception of green is itself a testimony to the success of packaging for food preservation. Mold is something the buyer of a name-brand, packaged item no longer fears. The real shift of the perception of green, however, came in the 1980s, with the introduction of ConAgra's Healthy Choice line of foods. This was the most successful brand introduction of the decade, and its strong use of green in packages throughout the entire line indicated that the conventional wisdom was badly out-of-date.

The varied meanings of colors are confusing when they are analyzed in isolation, though they are hardly confusing in everyday life. For example, Americans associate green with the color of their money, which connects with the deeper psychological connection of green with hope and with an inoffensive, general acceptability. Nearly everywhere, green is also associated with nature. In Western Europe, and to a much lesser extent in the United States, it is associated with a political movement that seeks to place protection of the natural world above economic values. "Buying green" is sometimes a political statement, sometimes an expression of interest in health and nutrition. The same colors can be used, and there is some overlap of products. But the consumer generally needn't think twice, or even once, about making the distinction. Americans may sometimes refer to dollars as lettuce, but there is little danger they will mistake one for the other.

Color is powerful, and subtle differences can make big emotional differences. Nearly everyone agrees with that. The question is how to make use of this phenomenon and the information it has generated.

The designer's response is often to more or less ignore it. If you accept that some color responses are instinctual, while others are deeply embedded in the culture, you would expect a skilled designer to be able to take these into account while designing a package. The designer's survey of what the competition is doing and what the packages that are likely to be near the new package look like should offer other insights about how people expect such products to be represented. Design is a process of synthesis. Designers may be aware of how researchers and other designers have analyzed color and other aspects of the package, but this can only serve as a background to the fundamental task of creating something new.

"There used to be all sorts of conventions about what colors are appropriate for what kinds of products," said Richard Gerstman. "They are all out of date, as one by one, successful packages are introduced that defy each of the rules."

Gerstman's firm is one of the few design companies that has its own research arm, and not surprisingly its studies have often zeroed in on the very concrete issue of what people expect from a package. One important issue that the company has studied through focus groups concerns precisely what people want to know about a product, and in what order. "We were doing a line of sandpaper, and we found out that the first thing people wanted to know is whether you were to use it on metal or wood," said Gerstman. This seems obvious, but his client wasn't making that information prominent, nor was the competition.

His experience affirms that, although much research is aimed at nonverbal communication, words can also be very powerful. One of the firm's more successful packages, for Maxell videotape, won a competitive advantage by describing the grades of tape as "general use" or "special event." The boxes were also color coded, and the overall expression high-tech, but the use of plain English to describe how the tape was made to be used was the true innovation of the package.

The most common failure of package designs is not so much that they fail to communicate, but that they communicate the wrong thing.

In this context, a work like that of Favre, which seems at first glance to be merely a stream-of-consciousness outpouring of contradictory emotions, can have a measure of value. It suggests a wide range of possible responses to the designer's choices, some of which the designer might not have thought about.

Designers working on packages to be sold internationally have long used checklists to screen out things that might be offensive. These tend to be crude lists of what to avoid. They counsel designers to stay away from white in Morocco, violet in Egypt, black in Greece. They are too general to be taken very seriously, but they can help weed out some bad design decisions before they are taken very far. And sometimes, a finished design can fall flat on its face because it has ignored some simple cultural stereotypes. For example, a California design firm's prominent use of purple — an Easter color in Catholic areas of Europe — in the graphic design system for Euro Disney provided fodder for those who argued that the theme park was an insensitive imposition on Europe, "a cultural Chernobyl."

The Nippon Institute of Color and Design Research method of arranging products is potentially more useful for the designer within the Japanese market. This method graphically reveals where expressive opportunities lie for package design. It's probably no news that color combinations identified as "elegant," "gorgeous," or "natural" aren't used for antifreeze. But there might be a spot on the chart, somewhere near the boundary line between "dynamic" and "casual," that could represent that product and break through the clutter of aggressively modern, masculine products. The mere fact that a promising visual approach has not been tried does not imply, however, that it will work.

In working for the Schilling and McCormick spice lines during the mid-1980s, Cheskin + Masten did something similar, on a conceptual level. Studies had shown that there is a direct correlation between the number of jars of spices on the shelf and years of marriage. People don't use up spices, and they don't throw them away, but they feel uneasy about it. "Freshness was an unexploited attribute in spices," said Alan Cutler, the firm's senior vice president. "Spices are also the only grocery item where people duplicate the store display in their homes, by buying many jars of the same brand." These insights led to several design recommendations. The redesigned jars have full-color labels that pictured the spice inside as a growing plant. When the jars are gathered together, the labels are intended to look gardenlike, not dusty. All the bottles have freshness seals, not so much to preserve the spice as to communicate the idea. The company also introduced a line of spices in smaller packages to suggest that they can be used up while the spices are still fresh. And the structure of the jars was complex and customized, in an effort to keep private brands from copying the look. "I'm still convinced," said Davis Masten in a 1993 interview, "that there is a market for single-serving packages of spices." That is one piece of advice the client did not take.

This kind of research concerns what Masten terms "life context" — how people live and how products and their packages fit into their lives. Some designers have long been concerned with this, at least in a visual way. For example, the designer Irv Koons kept up-to-date with trends in home decorating through suppliers' catalogs, paint colors, wallpaper, decorating magazines, and other materials that gave him clues on designs for Dixie cups, Scotties tissue dispensers, and gift boxes for liquor and cigars. Masten has spun off an affiliated firm, Imagenet, to do ethnographic research to determine how people are actually living and using the packages.

"Packages aren't only important in the stores," said Christopher Ireland, Imagenet's president. "People's shelves at home are a gold mine of information." The way the company gets into people's cupboards and closets is by paying people to take pictures of the way they live. The trick, Ireland says, is to give them way too much film and require that they use it all. Thus, after the relatively formal shots are finished, people start taking pictures of their bathrooms, their closets, their drawers. "You can see which products are on display and which are hidden," Ireland said. "You can see packages used for all sorts of things for which they were not intended — propping things up, holding them down. Kids are constantly shown eating out of packages. You hardly ever see a plate. You realize how much *worth* packages have."

Much of the work done by Imagenet is for youth-oriented clients, whose executives understand that they really don't understand young people. Ireland said one client, a shoe company, thought it might tap into a growing ecological consciousness by eliminating the boxes in which shoes were sold. The photographs taken by young women like those who were expected to buy the shoes indicated otherwise. They showed the shoe boxes are not thrown away but are frequently used as an organizing system in the closet. In other words, the box is part of the value of the product.

"I think that the coming of various kinds of shopping-at-home services might change packages, but, if anything, it will make them more important," Ireland said. "People at home are more aware of their environments. They'll look at a package and say, 'That won't match my wallpaper!' or 'Where in the world am I going to *store* that?' "

This kind of research potentially has as much impact on the kind of product that will be offered as the kind of package it will come in. But if half a century of package design research shows anything, it is that most people are unable to tell the difference.

Package design research is less often used for exploration and inspiration than as a form of insurance. The reason is obvious.

It costs millions of dollars to launch a new product. The great majority of new products — as many as 90 percent by some estimates — fail. Even extensions of existing product lines, which, like movie sequels, are often done because they are less risky, still fail far more often than they succeed. While as recently as thirty years ago many products were in the hands of strong-willed proprietors who viewed packaging as an expression of themselves, today nearly all are in the hands of managers driven by fear of a career-ending mistake. In such a dangerous atmosphere, it is understandable that executives would embrace research — both to give them a competitive edge or provide someone else to blame if things go wrong.

Executives do not always take researchers' advice, however. One designer, faced with modifying an extremely well-known package for a food product that people remembered fondly but did not buy, did several alternative designs. One was a radical departure that was popular with the young, another a conservative approach whose principal supporters were men more than forty-five years old. "The research was correct," he said. "The executives, all of them males over forty-five, loved the package, and they picked it." The modest repackaging helped sales, he said, though not necessarily as the research suggested the other alternatives would have.

One leading package design firm, Primo Angeli of San Francisco, has announced a novel approach to cutting the financial risk of introducing a new product. The firm will design packaging for proposed products that do not yet exist. The packaging can then be tested and the marketing concept refined. Only when it's clear that the company has a winner on its hands will it need to go to the expense of actually developing the product. This is the ultimate in Cheskin's principle of sensation transference. First you engineer the sensation and then you engineer the substance to which the sensation is to be transferred.

The assumption that underlies the Primo Angeli strategy is that engineering the sensation is less expensive than creating the product itself, but more likely to fail. Most people would probably find the approach logically backward, but while packagers believe that the product itself must at least be acceptable, they know that logic has little to do with what people buy. Certainly, if such a method catches on, it would depend heavily on life-context studies to generate ideas for possible new products.

At the moment, though, most package research is evaluative. It is used to test a proposed new design or as a way to help choose among several design alternatives. For example, a company considering a line extension might commission a study of possible type styles for the name of the product. Test subjects are shown the alternatives and are told that these are different products. Then they might be asked questions about which they think is the highest-quality brand, which they think represents the best value, which is worst, and which they would most like to receive a free supply of. This last question has been found to work better in determining what people will buy than asking whether they will buy it. Accepting a freebie implies less responsibility than making a purchase.

Such indirection is a fundamental premise of package design research. In almost every case, the researcher asserts that the product is what's being tested, not the package. If you ask about the package, most researchers and designers agree, people will look at it as an aesthetic object. And while the responses on that level might be either acute or stupid, they will certainly be irrelevant. Nobody goes to the store as if it were an art museum. The visual overload is too great. Besides, the point of going to the store is to buy things you need or want, not to have an aesthetic experience. People cope by not looking, and still they see plenty.

It might seem that the person who is confronted with the word "lite" in four different typefaces — and nothing to eat, drink, or even touch — would catch on. It seems pretty clear that what is being measured is not four different products but four different approaches to visual communication. Yet that doesn't seem to be the case. We are all so accustomed to looking past the package that we don't see it even when it is put right in front of our eyes.

Passive observation, either through a video camera or by an observer, can also provide a sense of how people react to a package design. It's also possible now to do eye tracking in a store, to see what catches the eye, what holds attention, and what gets avoided.

One of the oldest and most used devices in package research is the tachistoscope, which is essentially a slide projector with a very fast shutter over its lens that can restrict the viewing of an image to a tiny fraction of a second. The image might be of a store shelf, and the subjects are asked to list all the products they were able to see. In some cases, the image will include a proposed redesign that is being tested. If the subjects can recognize it, that will mean that the product's equity is being preserved. If more recognize it than see the existing package, it is pretty clear that a redesign is overdue.

The blink of recognition, measured by a tachistoscope, is not nearly as important as the snap of judgment, whose consequences will unfold over a long period of time. Moreover, there is evidence to suggest that the brain does not first recognize and then judge. Either they happen simultaneously, or they are two ways of describing what is, for the brain, a single action. Recognition can imply either a positive judgment or a negative one. Admittedly, any recognition is a start, because a package that can't be recognized will never mean anything. But most package design research has always stressed sheer shelf impact. This is because it is difficult to monitor and test the thoughts and emotions that come into play between the time the product is first seen and when it becomes part of the shopper's life context.

One who tries is Stan Gross, a marketing consultant to major corporations, who has developed a technique he calls mental mapping. His premise is that about 90 percent of people's thoughts are not consciously accessible, though what he calls the inner mind does respond to the symbolism of packages. "I can't ask you why you like a certain package," Gross says, "and you can't tell me."

This inner mind of which Gross speaks is not the primordial subconscious, but rather a semiconscious realm of wants and expectations that have been shaped through culture. Gross explains: " 'I'll let you know how things feel,' the inner mind says. 'If it's good, fine. If not, I'll tell the rational mind to figure out what's going on.' "

For Gross, the supermarket shelves are filled with as many

heroes, witches, children, monsters, and stories as bloody and magical as the forests were for the Brothers Grimm. The inner mind operates, Gross argues, in the form of myths and other sorts of stories, including many that are accessible through popular culture. "There aren't very many human needs, but people have a lot of wants, and these are tied indelibly to culture." He says that physiological tests indicate the power that a package can have to change people's emotional states, but that it doesn't tell how to understand the relationship between the package and people's feelings. "The package is not a silent salesman at all," says Gross. "It screams — but it screams to your inner mind."

Using an approach that's part Vienna, part Catskills, Gross works with groups of subjects — sometimes consumers and sometimes marketers and package planners within his client companies — and urges them to play games that he says help elicit responses that come from the inner mind. He does not use a large sample, and while the subjects involved are all people who are users or marketers of the kind of product being tested, there is not much of an effort to be demographically representative. They start out playing basketball, blowing bubbles, drawing with crayons, or engaging in other activities that pull their minds out of an analytic pattern into a playful state. The serious part of the study is introduced in an equally whimsical way. The subjects are formed into teams and given, for example, a series of different packages, and then Gross challenges them to tell stories. If this toothpaste were a person, they might be asked, how would you write its obituary? Or if this detergent were a movie, what would it be about?

The method seems strange, and it probably contains as much art as science. But Gross's firm stays very busy doing work for well-known clients. The odds against success for a new product introduction or a line extension are large, and Gross has shown some success in beating them.

For one brand of toothpaste, a couple of different obituaries pointed to a character that seemed familiar. He was born in Austria and became a bodybuilder, then a famous movie star who always triumphs over tremendous physical challenges. This was a character the subjects seemed to feel comfortable with, but for Gross, it meant trouble. "You might think Arnold Schwarzenegger is great, but that doesn't mean you want to have him in your mouth!" Gross says. He recommended some specific changes to the package in an effort to make the product seem less muscular and likely to go out of control.

Another toothpaste with blue packaging elicited associations with *The Little Mermaid,* presumably the Disney version. For Gross this was not simply a result of the color of the packaging, which might have been expected to produce such underwater associations. "This is someone who is trying to be something that she's not," Gross said. "This is a product that is making too many promises, and because it is attractive people might sample it. But they won't adopt it as their regular toothpaste because it seems like just a novelty."

He encourages a certain amount of outrageousness, because he thinks the ways in which people are playful reveals a certain truth. One very familiar brand of detergent, for example, tests as a real slut. She stayed out all night at fourteen, was pregnant at seventeen, and never told her mother. Her boyfriend's out of work, she drinks too much, and she recently attended a Halloween party wearing a loincloth and no top. The source of this was that she's "easy," promising too much to too many people. The suggestion that Gross found in the story was that people who wanted a product that claimed to have so many features were lazy, low-class people. The group at the next table, working with Tide, saw it as Sylvester Stallone. "The positive side of this was that it doesn't beg the question. It gets your clothes clean. It gets the job done," says Gross. "The negative is that the product is a little bit crass."

"We once did a logo for a line of new food products that knocked your eye out," says one package designer. "We thought it was great. The client really responded to it, and some research showed that it was highly recognizable. Then Stan Gross tested it — and it was *Fatal Attraction.*" On the strength of this reaction, the client decided to tone down the logo, and the designer now thinks that this was the right choice. "The earlier design was

231

just *too* seductive, and probably not appropriate for the product line."

Gross systemizes these responses into "Marketing Maps," Gross's trademarked term for large charts showing people's responses to the client's package and to the competing brands tested. They are organized so that comments that relate to similar issues are close together, and the strengths and weaknesses of the various packages can be seen. For example, one cinnamon-flavored gum is seen as skinny and active, another as an overweight underachiever. Such details from these imagined biographies are linked with judgments the subjects express when discussing the biographies. For example, the first gum's A grades in his high school shop and speech classes are connected to statements that the product is well made and has not been successfully imitated. The second gum's C in science indicates that it's not well made, and Cs in art and music indicate that the gum isn't refreshing. At the top of the chart are the big conclusions. The first gum "gives us a sense of reinvigoration and warmth, warming up our brains and body to be ready to get on the go and make achievements." The hapless competitor "is an ill-bred gum for ill-bred people who chew this gum."

This preliminary map becomes the basis for a final map that goes into greater detail about the client's product and contains substantial analysis, along with some specific recommendations that can concern such issues as the name of the product, the package design, or even the kind of package in which the product is dispensed.

For example, Gross credits himself with the introduction of the aseptic snack pack in the United States in 1984 by Knouse Foods, an applesauce producer. The technology had long been available, but a good reason to use it had not been identified. Gross placed some working mothers under hypnosis and found a good deal of guilt about what they packed in their children's lunch boxes. They saw applesauce as a healthier food than what they usually gave their children for dessert, but they knew that if they packed it in a Tupperware container, it wouldn't get eaten. "Applesauce in Tupperware is ammunition that's going to fly through the air — the bullies against the nerds," Gross said. "We thought that would be less true if you gave kids their own personal container, rather than

Mom's container." His client worked with Continental Container to develop such a package, and now, about a third of all applesauce is sold in snack packs.

The guilt of hurried mothers plays a big role in the packaging for a frozen-steak-sandwich product on which Gross advised. The package shows a sandwich piled with vividly colored tomatoes, peppers, and other vegetables, which suggest at least the possibility that the product can be part of a nutritious meal that is also festive and warm. "These colors speak to the consumer and say something they haven't heard before," he argues. "There are no rules that the red means this and the green means that. People don't pick apart the package; they see it as a whole."

This healthy-looking steak sandwich is a bit like the cracker box's "serving suggestion" that used soup to make the product appear to be nutritious. The difference is that the cracker box, by adding the nutritional value of the soup to that of the crackers, was trying to fool the intellect and had a pretty good chance of being caught.

The vegetable-laden steak sandwich, by contrast, does not imply that the package contains anything that's not there. What it does embody is a series of unspoken promises that the product can be used responsibly for the family and as a focus of activity. But if Mom does not find the time to make it into a beautiful, vegetable-laden presentation, and if the product is used as a quick expedient rather than as a focus for strengthened family feeling, nobody is going to blame the product. In analyzing the situation — that is, by shifting to a conscious, rational mode of thinking — it's clear that there is only so much you can expect from a chunk of thin-sliced frozen beef.

"We can't create a want that doesn't exist," says Gross, who does not accept clients that sell liquor or tobacco products or political candidates. "What we do is identify what the want is."

But Gross and other psychological marketers are dealing with fears and desires that go deeper than whiter laundry or something to chew on. Parents worry that they are failing their children. Individuals feel lonely and unfulfilled. People feel powerless to do anything to change their lives. To assuage such deep anxieties and psychic needs through such palliatives as a frozen steak sandwich, a pack of cinnamon chewing gum, or a more powerful laundry detergent seems cruelly inadequate. It seems as if life ought to offer more than this.

Gross argues that people know, on some level, that the purchases that they make will not fulfill their deepest wants. "Buying things is a way of coping," he says. "We all do it. Some go out and buy a box of Russell Stover candy. It's better to do that than cocaine." As the Scarecrow, the Tin Woodman, and the Cowardly Lion discovered, the Emerald City offers compensations for the deficiencies we feel in ourselves. We may see that they are empty symbols, but we pursue these bright baubles because life would simply be too dark without them.

Gross's methods demand that his subjects perceive the product and its package as a person. One of the major themes of the history of packaging is the replacement of salespeople — and the danger that they will communicate extraneous or undesirable messages. Gross's work seems to indicate that packages, far from being instruments of clear, positive, well-controlled communication, actually elicit emotions as complex as any human salesclerk. Indeed, many of the products Gross has studied — the bullying, presumptuous candy bar, the sexually licentious, unreliable laundry detergent, the dumb chewing gum that manages at once to be overweight and sugar free — are extremely negative. If they were real people, they'd have a tough time getting a job as a salesclerk, and they'd have difficulty holding on to a job. The vaunted "silent salesman" may not have salt mackerel and kerosene on his fingers, like the clerk in the old Uneeda biscuit advertisement, but he is still capable of scaring away the customer.

Yet there is something reassuring in the suggestion that even buying a packaged product off the shelf is a kind of human transaction. Human survival has always been in some measure dependent on being able to instantly judge whom to trust. Familiarity is one criterion. The familiar face can belong to a cousin you've grown up with, or to a package you've grown up with. The standards

234

people use to judge the stranger, or the new product, are elusive and difficult to express. Still, most people feel confident enough in their judgment to trust first impressions of the people they meet and behave accordingly. Those involved in package design and research often speak of successful packaging in terms of such qualities as clarity, integrity, a unity between outward appearance and inner qualities, unpretentiousness, and effectiveness. These are all personal qualities. We use packages, in part, to escape the anxiety of having to decide whether to trust other people. And still it is likely that many of the same instinctive and deep-seated cultural mechanisms come into play when we look at a can, a bottle, or a box as when we look into the eyes of a fellow human being.

From one point of view, this is a degradation of our humanity, the replacement of human relationships with empty, wasteful simulations. Addicted to shopping, people look to products to fill the voids in their own lives. And the makers of those products are becoming increasingly adept at tapping into and manipulating the primitive emotions that were useful for survival in the era before packaging.

What packages try to do is make people confident about things they do not understand. Sometimes this confidence is misplaced. Swaim's Panacea was, for its era, a great package, but the contents poisoned those who used it. It's no accident that the nineteenth century gave us the term "confidence man," a fraud who preys on desperation or greed. Nobody wants to be conned, and Gross's studies seem to indicate that people have strong defenses against what they perceive as manipulation. The trick, of course, is to manipulate people in ways they do not even recognize, let alone understand.

There is, however, another side to confidence. It is a prerequisite to technological and material progress. In the modern world, no technology really succeeds unless it passes from the hands of the engineers who understand it, and the mechanics and technicians who know how to fix it, to the larger public that can do neither. From domestic electricity to the automobile, radio and television, and the computer, this progression has been played out again and again. The role of design in these and most other areas of industry has been to create an aura of unearned confidence, to make the promise that something can be used and that it will make life better. You can, of course, wonder whether particular technological changes, or even all of them together, have really brought progress to human beings. But very few people would be willing to give up the greater mobility, comfort, access to information, and longer and healthier life span that technological society has brought about.

As we have seen, packaging by protecting and preserving products and by assuring standard quality has been an important, though little noted, part of technological progress. And packaging may, in fact, be the most successful area of modern design, precisely because it calls the least attention to itself. It is virtually identical to the product. By contrast, modern architecture, with its imitation of industrial machinery, is primitive. It is possible, even necessary, to talk about style as something separate from its content. The same has often been true of appliances and other forms of product design. Streamlining during the 1930s, and angular jet age styling during the postwar era, have encouraged users to see such products as refrigerators, vacuum cleaners, or pencil sharpeners as if they were something else. But while package design often incorporates some of the fashions of its time, its goal is most often to be the face of the product itself.

Package design is successful to the extent that it communicates very complex messages to people who give it little conscious thought. This may prevent packaging from becoming art, but it does define a unique strength that is potentially very valuable. Emotion can be seen as a very rapid means by which humans process information. And no field of design deals more effectively with the emotions than does packaging.

Consider, for example, computers and the communications advances implied by the term "information superhighway." In computer terms, a package is an interface, a device that tells you what it is and how to use it. Packages are generally better at doing this than most computer software. Those interfaces that are considered most user-friendly, such as the Macintosh operating system and Microsoft Windows, use picture-based icons. But as applications become more complex, the limitations of this approach have become evident. Icons have proliferated to the extent that they threaten to become a pictograph-based language, like Chinese, in which the characters lose their mnemonic value and must simply be memorized. This shows that, although the interfaces use pictures, their structure is linguistic.

By contrast, packages unashamedly use words where they are most appropriate, but they also deploy powerful visual messages that the brain is able to grasp without conscious analysis. If this kind of power could be applied to software, it would bring yet another revolution to the computer industry as computer use becomes truly universal, and scarcely conscious. Surely, navigating through the often promised world of five-hundred-channel cable television would be easier if people could do it in their accustomed supermarket trance rather than have to struggle with a road map. Indeed, the experience of the information superhighway will probably not be that of driving a car at high speed through a featureless landscape. Rather, it will be like moving at your own pace through an environment of vividly presented choices. The future will be like going shopping.

What then of Keats's beauty and truth? The short, easy answer is that a package is the wrong place to look for either. And yet it seems clear that a poet two millennia hence would certainly learn more about our world by looking at cans from a supermarket than canvases from a museum. Moreover, the aesthetic qualities of packages, because they are more closely tied to psychological perceptions of how people see and make judgments, would very likely be more intelligible over time than would the devices used by more individualistic fine artists. Thus there is a kind of truth and a kind of beauty to be found in packages.

Keats's urn was an object that showed life vividly, yet suspended in time. In *The Symposium,* Plato suggested that the desire for beauty grows from a desire for immortality. Packaging seeks to deny decay even more explicitly than did Keats's urn. We expect the contents of packages to be virginal, untouched by human hands. People lust for immortality, yearn for perfection, and packaging, in its mundane, limited way, responds.

9 *Empties*

Packages sometimes embody a different, horrifying kind of immortality. They are the Undead — whose lives are spent but who linger on to haunt and curse the living. No longer unravished brides, these are hags and zombies whose ugliness is as extreme as their one-time seductiveness. Empty and eternal, they are specters of regret, lurid reminders of past indulgence. You can discard them, but they do not disappear. You can bury them, but they do not decay. These immortal packages — the ketchup-smeared burger boxes, the souring yogurt cups, the plastic bottles that have been squeezed for the last time — are waiting to engulf us all.

Public-policy analysts talk about this problem in a less emotional way. They speak of MSW, an even less evocative term than its full name, "municipal solid waste." A close and dispassionate look at household refuse — something that most people are unwilling to take — gives a picture at odds with the horrific visions. Packaging that people discard personally in their homes accounts for only about 15 percent of solid wastes in the United States. Moreover, packaging wastes aren't growing nearly as rapidly as population and the number of households. And packages are losing weight and wasting fewer materials. Indeed, the thing that makes the municipal

238

solid-waste problem special is that, unlike so many other ills of modern society, it is being addressed with some success. (Toxic solid wastes are a different story, though the two issues are frequently confused.)

It's not surprising, though, that solid waste, and particularly used packaging, is an emotionally fraught subject. There is shame in elimination. People don't like to touch or even think about their wastes. Garbage must disappear, be kept out of sight. Paradoxically, societies rarely succeed in doing this because they are reluctant to commit resources and talent to garbage collection and disposal. Even in countries with well-developed systems of public services, solid waste is the one that is most often in danger of breaking down. People avert their eyes from the trash, but they know it's there. They don't want to think about it, but nevertheless feel a vague, unresolved responsibility.

When you add to these conflicted feelings the emotional content we have seen to be lurking in packaging, the combination is very powerful. Newspapers give solid waste much of its bulk, but packages give it its personality. They were charismatic in the store, and they draw the eye in the landfill as well. Packages, including corrugated boxes, wooden pallets, and other shipping materials most consumers never see, constitute about 30 percent of U.S. municipal solid waste. They are extremely visible in landfills, but they are even more prominent in the landfills of our minds. And plastics are viewed as the worst of all. Even though plastics account for about 13 percent of packaging discards, they get attention and stir passion. People imagine mountains of fast-food containers, whole ranges of brightly colored plastic bottles, dead but immortal, and landfills oozing across the landscape like a viscous horror movie slime monster. Zealous environmentalist teachers tell their kindergarten students that soft-drink containers kill cute, lovable seals. Grass-roots movements rise up to ban the multilayered, nonrecyclable juice carton. And the principle of punishing the packager seems to be gaining popularity throughout the world.

Opponents of packaging argue, with some justice, that such extreme reactions are merely necessary correctives for a culture long bombarded with the message that waste is not merely acceptable but essential to progress and prosperity. What this thinking has given us, they argue, are antimonuments like Staten Island's Fresh Kills landfill, which holds most of the wastes generated by New York City during the last half century. Fresh Kills is the second-largest human-made structure in the world, surpassed only by the Great Wall of China. It is an anomaly in its size, visibility, central location, and outmoded management methods, but it serves as a convenient and potent reminder of the waste inherent in modern life.

Nobody who has ever seen Fresh Kills or any of the immense landfills that now constitute the state of the art in American refuse disposal will deny that solid waste is something to be concerned about, and that packaging is part of the problem. The emotions that are inherent in packaging inevitably shape debate about what should be done. And the responsiveness of packagers to consumers means that this public-policy issue is being decided more at the checkout counter than through the political process. In such a situation, perceptions count for quite a lot.

This is also an area in which change is happening rapidly, as new packages, new government policies, new technologies and processes are introduced. Statistics on materials use and recycling are out-of-date by the time they are collected. What follows in this chapter isn't an attempt to give a comprehensive account of packaging and the solid-waste problem. Rather, it continues the earlier chapters' examination of packaging as a cultural phenomenon to the inevitable question: What happens when the package is empty?

There is probably no other environmental issue that engages more people than does solid waste. Recycling of packaging and paper has rapidly changed from a voluntary cause for activists to a routine for householders, often collected at curbside like their trash. Shoppers are encouraged to purchase products because of the environmental qualities of their packaging. Increasingly, they tell pollsters that this does make a difference, though not always a decisive one. Employment is rising in recycling industries, and the desire to save materials and ease recyclability is probably the single most potent force changing what consumers see on their store

240

shelves. Increasingly, packages are being shaped by environmental concerns.

In the United States, government involvement in the issue tends to be through the institution of recycling programs to help municipalities ease their waste disposal expenses. Elsewhere, especially in Europe, regulation of packaging is increasingly taking hold, along with restrictions that seek to induce major changes in lifestyles.

The European Union has set a standard of 50 percent of all wastes recycled by 1998, and some of its members have set even more demanding standards. The Netherlands has embarked on a program to ensure that no packaging waste at all will be placed in landfills by 2001. In Germany, consumers are required to leave secondary packaging, such as boxes that cosmetics and toothpaste come in, at the retailers' and bring all the primary packaging back to the store when it is empty. As an alternative, shoppers can purchase products with a special green-dot logo whose packaging can be deposited in special bins for curbside pickup. The purchase price reflects the recycling costs.

The workability of these programs has not yet been proved. And even if they work in the Netherlands and Germany, two countries with traditions of compliance and well-maintained public environments, there is no guarantee that they will work in the less cohesive and cooperative environment of the United States. Nor is it clear that a procedure required for a place like the land-poor Netherlands, where any hole you dig becomes a pond, is appropriate for a continental, relatively thinly settled country like the United States.

Such regulations are not simply differences in procedure, but major cultural changes that are sometimes not recognized by the bureaucrats that institute them. In Tokyo, a 1993 garbage regulation severely tested Japan's famous propensity for social cohesiveness and compliance with law. Households were ordered to place their refuse in clear plastic bags with their names on them, so that it would be easy to determine who wasn't recycling. This was not merely onerous, it was impolite. In Japan, wrapping is a sign of respect and discretion. Being forced to display your refuse in public would be an imposition. A revolt of wives, and the unavailability of the required bags, forced a delay.

What happens in Germany, Japan, Italy, Switzerland, and Great Britain, all of which are dense countries with solid-waste problems and sophisticated packaging industries, will affect the rest of the world. That's because companies operating in those countries will be forced to engineer new packages that work within those countries. And less wasteful, easier-to-recycle, and potentially less expensive packages will be attractive to product marketers throughout the world.

There have been many calls for government regulation of packaging in the United States. Legislation has been introduced in most states to require companies to make greater use of recycled content and severely cut the amount of packaging materials sent to landfills. A few such laws have passed, though most have been defeated. On a national level, there have been proposals for dramatic reductions of packaging discards, though at this time, they do not appear high on the political agenda. It seems most likely that Americans will deal with solid-waste issues largely through local recycling programs and the widespread expression of consumer preferences, sparked in part by the obligation to recycle. Such a market-based approach offers the advantages of responsiveness and flexibility. The greatest potential disadvantage is that there is little role in such a process for disinterested expertise. The very companies that are responding to consumer perceptions are also the ones with the greatest experience in shaping them. Yet the large consumer-products companies have not sought to minimize awareness of solid-waste problems. Instead, prodded by environmental activists, they are competing to be seen as environmentally sensitive.

Consumers see garbage, and particularly packaging waste, as a very serious problem. In a 1993 poll conducted by Opinion Research Corporation for Gerstman + Meyers, respondents were asked what they considered the "single most important environmental issue" and were not prompted with multiple choices. Consumer solid waste finished first, cited by 23 percent, followed by 20 percent for air quality, and 16 percent for water quality. In each of the previous five years this survey was taken, air quality was ranked first, but it has slowly receded in importance. Respondents who were most overwhelmingly concerned with consumer solid waste were those eighteen to twenty-four years old, who cited it twice as often as air quality, and twenty-five-to-thirty-four-year-olds, who cited it by an even more overwhelming margin. These are heavy-consuming demographic categories that are of particular interest to marketers. (Another 1993 survey, done for the Committee on Packaging and the Environment [COPE], an industry group, indicated that these younger consumers were also the least likely to recycle.)

These are numbers that are impossible for any maker of consumer products to ignore. Statistics like these underlie the proliferation of products on supermarket shelves that use recycled materials, are designed to be easily recyclable, or tout the fact that they use less packaging. Such changes are required more by the marketplace than by the environment. Companies' responsiveness to these perceptions helps validate the consumers' perceptions. It does not, however, indicate that they are correct.

Solid-waste problems are fundamentally different from those that relate to air quality. Releasing wastes into the atmosphere exposes everybody to substances that can threaten their well-being and even survival. It is possible to trace filthy air directly to bad health, lost human productivity, medical costs, and death.

Consumer solid waste creates no such health hazard. Unlike air pollution, to which nearly everyone is exposed, solid consumer refuse is, for the most part, buried in sanitary landfills where it affects hardly anyone. The landfills are lined to prevent contamination of groundwater, and they are compacted and covered each day with a layer of dirt to reduce vermin and unpleasant odors. This is a system that stops even normal biological degradation. The public would prefer that their garbage would rot and return to the earth. But the lack of such organic activity in landfills does at least have the advantage that carbon dioxide, which may contribute to global warming, is not being released. There is, of course, pollution associated with packaging, as there is with nearly every other economic activity. Cutting down on the wastefulness of packaging can contribute to the alleviation of many pollution problems. But the mere existence of huge landfills and the need for more of them is not a pollution problem.

Waste and pollution are linked. All materials processing and transportation create a measure of pollution, so if you use more things than you need, you generally create more pollution. But there are more direct ways of minimizing the air and water pollution involved in package manufacturing. And far more transportation-related air pollution is generated by thoughtless and unnecessary human trips than it is by moving boxes around.

Most of the passion that concerns packaging is really addressed to waste. Once we recognize that people die because of air pollution, but nobody dies because of household trash, we can start to put solid waste in its proper context.

It is a problem, to be sure, but it is more like letting old buildings deteriorate until they are useless or urban infrastructure decay because of lack of proper maintenance. All of these represent squandered riches, and it would be better to waste less of everything. Conserving buildings might actually contribute as strongly to solid-waste reduction as packaging reform. Garbage archaeologist William Rathje's studies of landfills suggest that demolition and construction debris is a more important component of landfills than packaging waste.

All biological and social systems generate leftover materials or energy. The difference between them is that biological systems are generally more efficient in reincorporating wastes into the system — recycling them. In most human situations, decisions concerning wastes are essentially economic. Poor people do not waste as much as rich people do because they cannot afford to, though they throw away just as much because they use a lot of secondhand goods that wear out quickly. Often, there are still-poorer people waiting to make use of the few things the poor can afford to throw away. The statistics on recycling for a city like Cairo, which has an entire class of outcast garbage pickers, would put most more developed cities to shame. People waste things because they feel that they can afford to, because they do not want to take the time and effort to make new use of their wastes.

244

In *Wasting Away*, his eloquent meditation on refuse, Kevin Lynch makes the fundamental point that the ultimate thing that people have to waste is their lives. Those who spend their lives in Cairo's garbage are trying to stay alive. Meanwhile, in Saudi Arabia, cars are simply discarded at the side of the road because their owners don't want to be bothered to fix them. It's easier to buy another car. In American cities, the affluent put quite a good quality of refuse out in the street, and the poor come along and go through the bags because it's worth their while. Waste is usually dealt with poorly because it is, first of all, a state of mind. It is something about which we do not want to be bothered.

When you consider the waste of materials as a trade-off against a waste of time, it forces a new view of packaging waste. Packaging has reduced the amount of time it takes people to cook meals, and the predictability and information it offers make for more efficient shopping. On the other hand, packaging probably makes people buy more than they need, which is potentially wasteful. But if we measure waste in terms of time, packaging is surely less wasteful than advertising. If Americans used the time they spend watching television advertising studying, learning new skills, or otherwise improving themselves, the productivity of the society and their own lives would undoubtedly rise. By contrast, packaging informs people about their shopping choices while taking up very little time.

The time dimension of waste has practical implications for waste reduction schemes. While there are some zealots who find great satisfaction in the categorization of their refuse for recycling, or in the creation and maintenance of their compost piles, most people don't want to take a lot of time. They balance their public-spiritedness against a sense of the value of their own effort. After a certain point, most people will inevitably see materials-conservation efforts as a waste of their time. And they will be correct.

The problems created by consumer solid waste are primarily financial and political. Tipping fees, the direct costs of placing garbage in landfills, have increased in recent years as the ability of governments and corporations to find suitable sites for landfills has diminished. In most of the United States, the difficulty of siting landfills comes not from a shortage of land but from people's unwillingness to allow one to be created nearby. In his book *In Defense of Garbage,* Judd H. Alexander calculated the volume of the holes Americans dig each year for such things as coal mines and gravel pits and compared it to the volume of solid waste generated. It turns out that the country digs up about twenty-three times as much land for resource exploitation as it fills with garbage, which could go into some of these big holes. The trend is toward fewer and larger landfills, located farther and farther from population centers. The public's unwillingness to live near its refuse is a direct cause of pollution — the contaminants that the trucks carrying the garbage release into the air. But this is not the kind of pollution that people think about when they consider the solid-waste problem.

About 15 percent of American solid waste is incinerated, a much smaller percentage than in Europe and Japan, where space for landfills is far more scarce. Up-to-date incinerators generate steam to make electricity from the garbage they burn, and in Japan such trash-to-energy incineration is considered to be a form of recycling. In the European Union some have argued for allowing incineration that generates energy to fulfill the new 50 percent recycling requirement. Adopting this definition would probably make it possible to achieve this difficult goal. To do so seems illogical, but it is true that all recycling has environmental costs. Although new incinerators are claimed to be clean burning and nonpolluting, this depends on excellent management and maintenance of the facility. The American public, instinctively understanding that refuse usually suffers from slack management, is less trusting. Again, however, any threat posed would come in the form of air pollution.

Efforts by incinerator manufacturers and operators may, in fact, be partly responsible for public perceptions of a nationwide garbage crisis. During the mid-1980s, the solid-waste-disposal business became steadily more concentrated, tipping fees increased, and some large cities and counties worried whether they would be able to find any place to put their trash. Trash-to-steam plants were presented as an attractive alternative to landfills, often by the same companies that had made the landfills more expensive. Several 1993 *Wall Street Journal* articles suggested that this garbage crisis was created by incinerator interests that needed long-term financial commitments to make their projects viable. American trash-to-steam plants have never become economically competitive with landfills. Cities that committed themselves to them are assured of a place to put their trash, something that seemed attractive when a landfill shortage loomed. But in most of the country the landfill crisis never happened. Meanwhile those localities that committed to the incinerators are paying higher tipping fees than those that did not do so, and in some cases, they are responsible for paying the operating deficits of incinerators running far below expected capacity. The apparent end of this phase of the solid-waste crisis has, in general, received less publicity than the earlier claims of a crisis.

That doesn't mean that there is no problem. In some highly populated areas, there is literally no place to put landfills. The trash problems of northern New Jersey may have more in common with those of the Netherlands than they do with Utah's. Moreover, landfills and incinerators tend to be built near the poor and the powerless. Waste-to-steam plants go to poor urban areas, while the latest trend in landfills is to put them on Indian reservations. In the United States, as in Cairo and Calcutta, trash is a very powerful weapon to make sure that society's outcast groups know their place.

And although the high piles of trash in landfills are not directly hazardous in the way that dirty air and dirty water are, there is some real pollution associated with packaging waste.

Perhaps most important is something that many do not recognize as a form of pollution at all: litter. A pollutant is a substance in a place where it can do harm. Litter is not a biological pollutant but a social one. Candy wrappers and fast-food hamburger clamshells thrown on sidewalks or roadsides may not cause diseases, but they are nevertheless injurious.

Litter is an insult to the people who live, work, or pass through the places where it is thrown. It is a marker of how little people identify with public space, a statement of how little regard they have for the people to whom it "belongs." The affluent litter on places that they perceive to be poor and powerless. The poor litter almost everywhere, because they feel little stake in the society as a whole. A little bit of litter begets more litter, and more litter draws lots and lots of litter. At its worst, in many poor urban and rural areas, this combination of lack of commitment and contempt leads to the ultimate insult — the dumping on sidewalks and vacant lots of truckloads of debris by people who were paid to haul it away but don't want to pay tipping fees. In some cases, the illegally dumped trash would not be accepted at many landfills because it includes hazardous materials. Packaging is not a major component of trash dumped in this way. Still, litter, which consists largely of packaging, tends to be a precursor to large-scale, illegal waste disposal. In most places, littering does not have such disastrous consequences. But it is still deeply disheartening to walk through an urban windstorm of sandwich wrappers, Styrofoam cups, and plastic bags.

It may be true that packages don't litter, people do. But that doesn't mean much. Packaging is not separate from the culture at large, but an integral part of it. Indeed, the central role of packaging in American culture has been to replace human relationships, which are ambiguous, time-consuming, unpredictable, and emotionally taxing, with expressive but less demanding containers. Self-service merchandising places self-satisfaction paramount and weakens individuals' consciousness of the society as a whole. Thus packaging can be understood not simply as the content of litter but as a part of its cause. And when the package becomes a kind of passive weapon, or an expression of despair, those who make and disseminate packages are inevitably involved.

Unfortunately, the nature of the problem is often misunderstood. Take, for example, the issue of fast-food packaging. As we have seen, standardized, disposable wrappers and containers are essential to maintaining the identity, standards, quantity control, and operating economy of the fast-food-restaurant industry. Such containers do not just eliminate the space-wasting, labor-intensive activity of dishwashing. They eliminate most other decisions and enable an unskilled staff to produce a uniform product of predictable quality. They are also convenient for customers, especially the

60 to 70 percent who carry their food out of the restaurant. But fast-food packaging is one of the most pervasive, visible components of litter.

Rathje has reported that fast-food packaging is a relatively small presence in landfills, about one-half of a percent by volume. But very few people actually examine landfills. They look at litter and see it not as a social problem in itself, but as part of a larger crisis of solid-waste management and pollution. They conclude that since fast-food packaging looms so large in the litter, it must be a big environmental problem. A 1989 Roper Organization poll found that 67 percent of consumers viewed fast-food packaging as wasteful, the worst showing of any product category.

McDonald's, the largest fast-food company, responded to the problem as primarily an environmental one. After exploring a foam recycling program, the company hired Dr. Richard A. Dennison, senior scientist of the Environmental Defense Fund, to conduct a review of its packaging. The major outcome was that the polystyrene-foam clamshell was scrapped as a container for hamburgers and replaced by plastic-coated paper wrappers. The new wrappers are not recyclable, but he predicted that the program would result in a 70 percent reduction of landfill wastes and a 90 percent reduction in the number of shipping boxes. Flat pieces of coated paper take up less space than molded foam both in shipping and in landfills, so if what you are worried about is mountains of trash, the paper is a better solution. He found that the environmental consequences of cardboard and foam cups for hot drinks are about the same, and the company stayed with the foam because it keeps customers from burning their fingers. McDonald's turned the criticism into a public relations benefit, and in the process did achieve a reduction of waste volume. It may even be true that paper wrappers are a less offensive form of litter than the foam clamshells, which are far more likely to be airborne. However, the public's failure to identify litter as a problem in itself and address it led McDonald's to undertake a program that, while worthwhile, will not have the results people had hoped for.

A more focused approach to the litter problem can be seen in the laws of several states that require that consumers pay a deposit on beverage cans and bottles, which will be refunded when the containers are returned to a redemption center. Such programs are based on the assumption that even if people do not feel any commitment to the streets, highways, parks, or forests, a small financial stake in the container will stop them from littering. Deposit programs have been quite successful in this goal, though the approach does create a management headache for retailers.

It is difficult to see how a deposit-and-return program could be more broadly applied to eliminate litter's social pollution. Requiring deposits on fast food would change the economics and operational procedures of the industry. In an American context, an onerous, tightly controlled program of solid-waste recovery, like those being tried in Europe, would very likely result in an increase in litter and illegal dumping, as retailers, on whom the primary burden for recycling falls, would be tempted to take shortcuts. Imposing special taxes on makers of products whose packaging is frequently part of litter might be a possibility, though it would be difficult to levy fairly.

You might assume that because so-called bottle laws gather containers at a central location they would be seen as a boon to recycling. In fact, many recycling advocates do not like bottle laws, because they take some of the highest-value recyclables — aluminum cans — out of the system. In a broader recycling effort, the high value of these cans helps pay for the recycling of some materials that are money losers. In this view, a somewhat effective litter-control strategy blocks a far more important effort to reduce material waste in the society at large.

Americans like to recast their social problems as technological ones, on which science and capital can be brought to bear. That doesn't work with litter, a product of the problems we have living with one another. Thus, this intractable form of social pollution is redefined as a symbol of solid-waste-disposal issues, which, though less serious, are better suited to technological answers.

Most of the other forms of actual pollution created by packaging are the consequence of its manufacture and distribution. Paper mills, for example, dirty an enormous quantity of water. They create foul-smelling air locally, and the burning of fossil fuels to power them contributes to more widespread air-quality problems while emitting greenhouse gases. Even paper recycling causes water and air pollution, though generally in less quantity than does the production of paper directly from trees.

Plastics production requires petroleum, both as the source material and as energy, and it must bear part of the environmental burden of the entire petrochemical industry. Petroleum consumption and its associated pollution by plastics is tiny compared with that used by personal automobiles, but it is nevertheless measurable and significant.

The enormous amount of energy consumed by the making of aluminum beverage cans is often excused because most of the power is generated by hydroelectric plants in the Northwest, which produce no emissions. However, these plants have entirely changed the ecosystem of the Columbia River, eliminating some salmon fisheries, a significant environmental consequence. Recycling aluminum cans requires only about 5 percent of the energy required to make them new from bauxite ore. That means aluminum cans are the most economically attractive recyclable material, and although their value rises and falls like that of any other commodity, they are likely to remain so. Still, they bear a substantial amount of original sin from their initial manufacture and must be recycled several times before they can be considered environmentally friendly containers.

Moving packages around creates pollution, too. The more packages you can get on a truck, the less pollution each one will cause. Thus, even though glass bottles are viewed as a somehow "natural" material, their weight and the materials that are used to minimize breakage often make them less environmentally sound than their plastic competitors. Such considerations are helping to determine the shapes of packages, as new structural designs are increasingly dictated by the ability to maximize the number of units shipped on a pallet and minimize their weight.

Obviously, the transportation component of pollution depends greatly on how far you are going to take the product. For a product distributed in a small area, a returnable, refillable glass bottle might be as "green" a solution as it seems. But for products that are shipped for long distances, a plastic bottle might cause the least damage to the environment.

In a system that has as many variables as the packaging, distribution, and retailing industries, there is no single way to guarantee that the pollution associated with containers is minimized. Indeed, it is possible to derive wholly contradictory results depending on the assumptions you make. Each of the different parts of the packaging industry — paperboard, glass, plastic, steel, aluminum — can make a case for its own environmental virtues and cite advantages over its competitors.

Recently, the U.S. Environmental Protection Agency has been promulgating a method called Life Cycle Analysis to try to provide a common structure for thinking about the environmental consequences of products. Essentially, it calls for the initial design of the product to consider the product's life beyond the point of purchase. The premise is that the buyer is only the immediate customer. The environment as a whole is also a kind of customer whose needs must be taken into consideration. The conventional marketing process is part of the picture. There's no point in making a lot of products that reduce materials and are reusable and recyclable if they are also useless to their immediate consumers.

But the approach also takes into account such matters as the existence of infrastructure for reusing or recycling and the embodied energy of the product, if it is to be disposed of by incineration. Questions about whether the raw material is renewable, like paper, or nonrenewable, like the petroleum used in plastics, must be balanced by calculations of the amount of nonrenewable energy sources expended in processing the renewable material. In 1993 the agency published a detailed set of guidelines, charts, and matrices that were intended to get everyone on a similar conceptual footing in thinking about the issue.

Life Cycle Analysis is similar to a concept promoted decades ago by the visionary engineer Buckminster Fuller. He called it "cosmic accounting," a term whose boundlessness suggests the difficulty of actually being able to do it. In fact, the boundaries that one places on the analysis largely determine the outcome. For example, several computer companies have won a lot of attention for their environmentally sensitive packaging. What if they had put the same amount of attention and money into reducing the electricity used by their products over their lifetimes? The benefits to the environment might well have been greater. Computer manufacturing involves some polluting processes, many of which have been moved to Asia and do not affect most of the customers. How do you balance the health of workers in Malaysia against the impact on American landfills?

Life Cycle Analysis is intended to apply to products, not simply to packaging. And if you view packaging not as an entity in itself, but as part of the product, the packaging can actually serve as a tool for lessening environmental impact. This is particularly true for food. If a smaller percentage of food production spoils because of protective packaging, the amount of food produced for each unit of energy expended (and air pollution created) is greater. Capital-intensive American farming gets more than twice as much food per unit of fossil fuels than does labor-intensive China, largely because of packaging that keeps food from spoiling on the way to the consumer. Such analysis demonstrates that packaging is not always a cost to the environment, but can actually be a major benefit. The challenge is to determine what kind of packaging is best for which products and how it should fit into the production, waste disposal, and materials-recovery systems.

Life Cycle Analysis demands a kind of knowledge of how people use materials that often does not exist. " 'Paper or plastic?' is the ultimate philosophical question of our time," Peter Cocatas, a consultant who specializes in Life Cycle Analysis told the 1993 Westpack conference and trade show in Anaheim, California. "The problem is to determine how many plastic bags it takes to equal one paper bag. Some say one, some say two, some say three. If you say five, a paper bag is unquestionably better, but if you say one and a half plastic bags to one paper bag, plastic will look very good." He said that he had not yet encountered an example of a material or a process that is superior to another in every situation. It's all very, very complicated.

*　　　*　　　*

And these perplexities are the last thing the public wants to hear about. "Solid waste is the one environmental problem that people believe they can do anything about," said Edward J. Stana, executive director of the Committee on Packaging and the Environment. His group's 1993 poll showed that respondents did not view trash and garbage disposal as the most serious environmental problem. Indeed, it was well below such issues as hazardous-waste disposal, water pollution, air pollution, and ozone depletion. (Unlike respondents to the Gerstman + Meyers poll, they were prompted with choices and allowed to cite more than one.) But when asked which environmental problem is the easiest to solve, they placed trash and garbage disposal first and littering second, with all the other problems far below. And by a large margin, the respondents cited recycling as the best way in which the problem could be solved.

Americans' compliance with recycling programs, particularly those that pick up materials at curbside, has exceeded all projections and in many cases overwhelmed the market for recyclables. There may be a latent expression of community spirit in this effort, and also a desire to assuage guilt about buying and using too much. It is certainly a more satisfying and often more practical thing to do than, for example, driving less in order to lessen air pollution. With recycling you feel that you can see the result.

So far, good intentions have not been accompanied by the kind of knowledge that would make consumer-recycling efforts truly efficient. About 20 percent of the materials collected by any recycling program are wrongly classified. This is a serious matter. One PET bottle, the kind that soft drinks come in, can contaminate twenty thousand high-density polyethylene bottles (HDP), the kind that are used for milk. To make matters worse, some similar-appearing packages are actually composed of different resins. A standard code has been adopted and appears somewhere on most plastic packages. But disposing of the trash is an activity that breeds inattention. "We say that we take plastic bottles," says Caroline Rennie, of Envirothene, a California recycler of HDP containers. "What we get are dead dogs and swimming pool liners."

She said her firm must sort the materials it collects at least twice before they are flaked and remelted into pellets that can be sold to create new products. Each sorting raises the price that must be charged and lowers the recycled materials' ability to compete in price with virgin materials.

Economic downturns, which lower consumption, depress prices for virgin materials, while the cost of recycling products remains the same, and supplies of raw materials are less predictable. Dependable markets that specifically require recycled materials have been expanding, but only gradually. That has inhibited the development and deployment of automated sorting equipment, especially for such lightweight, difficult materials as plastic. The mechanical plastic sorters that have been developed to date pass the trash through a complex Rube Goldberg series of obstacles and produce a product that is not high-grade. It is a dull material, somewhat greasy to the touch, used in park benches and garden edging. Most likely, the complexity can be reduced only by standardizing the composition and character of the packages in the first place, which carries the danger of preventing potentially more efficient innovations. Sorting a garbage heap will always be a laborious process. Moreover, the COPE poll indicated that only 19 percent of the public expect to pay more for products that come in recycled materials, while 48 percent expect to pay about the same, and 30 percent expect to pay less. That means that, despite public enthusiasm for recycling and a declared willingness to purchase products that are environmentally responsible, marketers cannot, at least for now, afford to charge more for greener packages. Recycled products still bear a bit of the stigma of having been used and discarded. They lose the magic of being "untouched by human hands."

Nevertheless, because of the participatory dimension of recycling, and its reassuring concreteness, recycling is widely seen as a virtuous activity fully compatible with contemporary, high-consumption lifestyles. That offers a marketing opportunity for many products. For example, cereal boxes, which have traditionally been made from paperboard with substantial recycled content, can now brag about this fact. Cereal boxes, which are conceived more as billboards than as containers, use a large amount of material to hold a relatively small volume. Changing the shape to a more efficient one or even changing the materials might have a greater impact on the environment and on landfills. But this is not widely understood.

Use of recycled materials, or even the assurance of recyclability, is an easily understood virtue. And recycling has been steadily on the rise for two and a half decades. According to a study done for the Environmental Protection Agency, about 26.2 percent of all used packaging was recovered from the waste stream in 1990, either to be recycled or to be reused in the same form. In 1970 the figure was only 9 percent. Aluminum cans were the most recovered items, at 63.2 percent, while 37 percent of all paper products were recovered. About half of all corrugated cartons were collected and recycled. This is a very efficient form of recycling because the cartons can be collected in quantity from central locations. Only 3.7 percent of plastic packages were recycled, but this figure is biased somewhat by the study's decision to measure the materials by weight. In conserving landfill capacity, volume, not weight, is what counts, and plastic recycling probably helps more than the figures suggest. While the economics of recycling suffer periodic setbacks, the overall trend is toward more, at least in part because recycling has such a positive marketing image.

It's not generally recognized that package manufacturers have always recycled substantial amounts of materials. For some processes, such as glassmaking, use of scrap material is essential to the process. However, most of the recycling consists of using scraps, leftover and other materials that have either never left the processing plant or that are discarded in central locations. The pollution-reduction payoff from this is high, because little energy is expended in collecting and transporting the scrap, and there is little chance of contamination. But this is often not considered to be "real" recycling, because it is simply part of the production process and because it does not reduce the amount of solid waste being landfilled or incinerated. That's why environmental advocates and manufacturers are increasingly measuring their recycled materials in terms of "postconsumer" waste. That means the stuff that people have used and discarded that has been diverted from landfills back into the production cycle. Postconsumer waste is far more expensive to process, and it inevitably involves higher energy costs. Thus, it is far more of a mixed blessing than companies' routine internal recycling, but it is the only kind of recycling that can actually reduce the volume of materials society sends to the dump.

Some advocate an even more rigorous standard, known as closed-loop recycling, in which used packages are converted into precisely what they were before. Thus, used PET soft-drink bottles do not get made into indoor-outdoor carpeting, as happens now. Instead, they become the bottles you'll drink from next month. This is a psychologically satisfying form of recycling, implying as it does that consumption can be without waste. This is, of course, an illusion. While it is true that some forms of closed-loop recycling, notably that of aluminum beverage cans, are practical and efficient, most cannot be. The energy — and hence the pollution — required to process discards into precisely what they were before is far greater than that of processing them into something that is different, but still useful. Moreover, with most materials, there is an optimum level of recycled content, after which the process becomes far less efficient. It is wasteful to try to segregate the materials used in packaging from those used elsewhere in the economy. In the natural world, wastes often become the key resource in a wholly different process. Similarly, recycled materials should be employed wherever they are useful. To argue otherwise is to expiate guilt about waste and consumption rather than to do what's best for the environment.

Recycling, although it is the best-known and most accepted way of dealing with package waste, is not always the best way to cut pollution, conserve materials and energy, and slow the growth of landfills. In most cases, the first choice is to use less stuff to begin with, a procedure that goes under the unwieldy, and according to the COPE poll utterly uncomprehended, term "source reduction."

People in the packaging industry speak of the three Rs of solid-waste reduction: "Reduce, reuse, recycle," and that is the order in which they are supposed to be considered. Reducing the amount of materials expended in a package means there is less potential waste. Reusing the package without reprocessing saves energy. Recycling uses more energy, but it at least assures that the package will be kept out of the landfill. Thus, what is supposed to be the third choice is perceived by the public as the best one.

Source reduction does have a very powerful group of allies — the corporate bean counters and number crunchers whose goal is to bring expenses down. In a sense, source reduction is one of the constant themes of the entire history of packaging. Cans, for example, have evolved from iron-walled barrels, opened with hammer and chisel, to very thin-walled membranes that in some cases hold their shape only because of the pressure of what's inside. Each generation has found ways to reduce the thickness of the metal walls and the amount of tin required.

In the early 1960s, a gallon of Clorox bleach was in an amber glass bottle that weighed 3.5 pounds empty and that had severe breakage problems. The first plastic bottles weighed 135 grams, a weight reduction of about 96 percent. But because the quantities are so large, with billions of bottles sold each year, it is highly worthwhile to continue to pare away each excess gram. Now the gallon Clorox bottle is down to 87 grams in the winter and 104 grams in the summer when the heat softens plastic and the bottle needs more strength in order to be stacked for shipping. Clorox, which makes a host of other products besides bleach, is moving toward absolutely minimal materials. It is able to count the weight of the materials it has saved and the energy costs of its shipment in the millions of pounds.

The Clorox bottle also contains 20 percent recycled plastic, which is mentioned on the product's label. The thinning of the bottle, which is at least as significant, is not mentioned, but it does contribute both to environmental improvement and to the company's bottom line.

The use of only 20 percent recycled plastic, as opposed to more, also illustrates the complexity of environmental trade-offs. In general, use of recycled materials entails some sacrifice of strength or resiliency. Depending on the use, packages with a large quantity of recycled materials must use more material, and their environmental advantage dissipates dramatically.

In some cases, the results of the thinning of packages are

easy to see. If you compare a contemporary PET soda bottle with a similar one from only five years ago, the difference is noticeable. The earlier bottle was quite rigid and required some force to be dented. The contemporary version is much more flexible, and although it's quite easy to distort the shape, dents do not remain. Glass bottles have competed with plastic by becoming lighter and thinner. And in some cases, rigid glass and plastic bottles have given way to flexible pouches that use even less material.

This trend has not made life any easier for recyclers, who must now collect even more packages and expend more labor in processing them to end up with the same quantity of salable materials. And it has probably not been widely perceived by the public, which still sees the same number of bright, expressive packages. But besides reducing the use of materials and the energy required to transport them, this light-weighting trend has an impact on the packages dumped in landfills. They are more crushable and thus take up less space. Even if the same number of packages are dumped, the mountains of waste will grow more slowly.

Other forms of source reduction are visible to shoppers. For example, many products, especially personal-care items such as shampoo and deodorant, are no longer sold with a paperboard carton protecting the primary package. Often, however, the inner package has to be redesigned so that it is more noticeable in the store, and for some products removal of cartons has meant that plastic containers have replaced breakable glass ones, a move consumers sometimes perceive negatively.

A recent award-winning package achieved source reduction by taking an opposite approach. A new product to break up drain clogs, introduced by Clorox's Liquid-Plumr brand, places a thin-walled plastic bottle inside a recycled paperboard sleeve. The sleeve reinforces the bottle and makes it easier to ship at the same time it provides a medium for vivid graphics to call attention to the product and show how it should be used and what its dangers are.

The other major source-reduction strategy involves the sale of goods, especially detergents, in concentrated form. This permits enormous savings in the use of cardboard and plastic, though at the outset, at least, it was also used to disguise a rise in the price per load of laundry. Brands that once placed whales on their packages, to dramatize their size, now run commercials arguing that their tiny containers express superconcentrated cleaning power. Along with a reduction in the overall size of the detergent box has come acceptance of a new compact shape, which moves away from the cereal-box billboard configuration in order to make a more efficient use of paperboard.

A variation of this approach is to sell refills of such products as cleaning agents and shampoos in flexible pouches and other demonstrably less expensive forms of packaging without pumps or other convenience devices. In general, these refill containers are difficult to use without spilling, and there is some evidence that people are using the refill packages as the primary container. The refill concept has, however, been understood and accepted by consumers.

This was demonstrated in an unlikely way in the marketing of a concentrated version of Ocean Spray cranberry juice cocktail. In a move similar to that of the detergent makers, this product, which is made from concentrate and was always marketed fully diluted, was brought out in a concentrated form. This saves not only packaging but also the costs of shipping water. Consumers, however, were not drawn to a concentrated cranberry cocktail. But behavior changed when the same package was relabeled a "refill." Shoppers understood this, and sales began to rise.

You can actually see packaging itself as a form of source reduction, especially for food. It is obvious that a can or a frozen carton of corn kernels contributes considerably less to municipal solid waste than would the same amount of corn sold on leaf-covered ears. When the food is processed in a central plant, it is possible to find uses for by-products that are not available to the householder. Corn is admittedly an extreme case, but Rathje reports that the amount of solid waste generated per person in Mexico City is higher than in several U.S. cities. The reason for this is largely the greater use of unprocessed foods, which leads to a larger amount of organic waste and simple spoilage. It is true that these natural products can degrade through biological processes, while a plastic pouch cannot. That is an advantage if you are composting your trash, which, given the thirty-one million tons of yard wastes discarded in the United States each year, wouldn't be a bad idea. In a landfill, however, biological decay matters little because landfills are operated in a way that inhibits natural degradation. There is no practical difference between a corncob and a plastic shampoo bottle, except that the shampoo bottle will probably take up less space in the landfill because it is easier to crush.

Although source reduction is a very effective cost-cutting tool for companies, and may be the single most effective waste-cutting technique as well, it does not get the positive consumer attention that companies want. It's hard to call attention to something that isn't there. When a package proclaims "less packaging," you can't help feeling something strange is going on — even if it's the right thing.

That's one reason companies such as Procter & Gamble that have quietly led the way in source reduction are far more vocal about their recycling efforts. When a giant like Procter & Gamble introduces a product in a recycled plastic container, as it did with Downy fabric softener, it can suddenly make the recycling of plastic a more predictable industry that justifies long-term investment. Recycling becomes part of the system, and major plastics manufacturers like Dow Chemical add recycled-content resins to their product lines. Recycling companies can borrow money for the long term, because they know that the next downturn in virgin-materials prices won't wipe them out.

In general, companies source-reduce for their own financial benefit and to anticipate and stave off possible government regulations. They recycle to improve their public profile as companies that are sensitive to their customers' concerns about the environment. The combination of the two is leading not only to a substantial reduction of packaging waste but to a transformation of the packaging-materials and recycling industries.

There is another reason that recycling is more attractive for product manufacturers to promote than is source reduction. Getting by with less stuff is, after all, quite a subversive concept. If people thought about it for very long, they might decide to get by with fewer cleaning agents, fewer cosmetics, fewer processed foods, and less of all the hope and promise that comes in packages.

The moment when the bottle is drained, the tube is spent, the box is empty, or the jar has but a few inaccessible blobs remaining is a crucial moment of the consumption process. It can be a time when you say to yourself, "I need some more!" But it can also be a time when you regret what you have done, when you feel guilty about the product and angry at yourself for indulging. It can be the end of an affair, something to be sought once more or avoided at all costs. Davis Masten's model of consumer experience labels this contemplative moment "resolution," the time when you decide whether this product is going to be part of your life or not. It is also the time when you have to figure out what to do with the empty package. And this more immediate, practical concern can help consumers take their minds off any guilt or regrets they might have about using the product. Recycling keeps the consumption engine turning.

While consumers' perceptions and feelings about the impact of packaging on the environment may not be wholly reliable, it is clear that packages must be designed to be discarded easily and without guilt. The refillable containers from the Body Shop constitute a participatory approach, which not so incidentally keeps customers coming back into the store while shunning cosmetics discounters. Aseptic juice cartons, which are not recyclable but waste fewer materials and use less energy than most of their competitors, have a harder time making people feel good about having used them.

The most poorly resolved package structure in recent history was surely the so-called long-box in which, until 1993, compact discs were marketed in the United States. While these plastic- and paperboard-wrapped repositories of mostly empty space were hardly friendly to the environment, they provoked a public outcry largely because they were psychologically disastrous. Disc buying is a somewhat addictive and indulgent activity to begin with. The long-box forced purchasers to deal immediately with a bulky, difficult-topenetrate package, all of which was waste. People hoping, through music, to purchase a measure of transcendence were left instead with a dispiriting pile of trash.

How did this happen? To begin with, the $4\frac{7}{8}$-inch-diameter discs have gradually replaced the 12-inch-diameter long-playing records, and this shrinkage inevitably led to a shrinkage of selling space on the package. The cardboard sleeve of the LP seemed a perfectly appropriate package for the product at the same time as it lent itself to graphics and notes that were full-fledged cultural expressions in themselves. The CD, less than a quarter the size of the LP, is presented as a precious miniature in a permanent, clear plastic, so-called jewel-box case. The jewel-box case, though it is prone to cracking and hinge problems, does protect the disc and create the sense that this is a luxury product. (Some record stores sell empty jewel boxes for as much as a dollar apiece, a practice that may increase the perception that buyers are paying for the package rather than the music.) The jewel box is also intended to last as long as its contents, so it is not perceived as wasteful.

Still, there were doubts about whether the 5-by-$5\frac{5}{8}$-inch jewel box could carry enough information and emotional content to close the sale. Moreover, retailers — the first constituency packaging must please — had problems of their own. Their stores were set up for the sale of LPs, and many were reluctant to invest too much to carry this new product. They remembered eight-track tapes and other failed formats. Moreover, the small size of the CD increased concerns about shoplifting.

In most of the world, the discs were sold in jewel boxes wrapped in plastic film, but the record companies' solution for the American market was to place the jewel boxes inside plastic or cardboard long-boxes, about twelve inches high. This provided about half as much selling surface on the package as on an LP, allowed merchants to refit their stores simply by dividing their LP bins in half, and created something that was harder to steal than the jewel box. As soon as they got the package home, consumers had to attack it with scissors and throw nearly all of the material away. In a few cases, record companies designed the long-box as a collectible. But it was difficult to remove the jewel box without destroying the long-box, and unlike an LP sleeve, the long-box had no continuing usefulness.

While CD packaging is a negligible portion of solid waste, both by weight and volume, this became an important symbolic issue. There is something irritating about an artistic expression causing so much garbage, and a few prominent recording artists made a fuss. Some designers created highly inventive cardboard packages that served as long-boxes in the store but could be folded to jewel-box size once they were taken home. These replaced the high-quality plastic of the jewel box with less expensive cardboard and also won very favorable reviews for returning art to recording packaging. A more modest approach, called the EcoPak, has plastic innards but a paperboard exterior that can carry far more vivid graphics than can the paper inserts used in jewel boxes. These EcoPaks are currently used for computer CD-ROMs, but all the major recording companies agreed in 1992 to sell their products in plastic jewel boxes without long-boxes. There has even been some source reduction, as increasingly two-disc sets are sold in jewel boxes only as thick as those previously used for only one disc. The record companies' agreement solved most of the environmental and psychological problems, though it failed to satisfy those who believe that packaging for music should itself be sensual and expressive.

It may be that the record companies did not make the best possible choice. But it is significant that they did make a choice, that they made it relatively quickly, and that by the summer of 1993, record store employees from coast to coast were tearing off every long-box in their bins and remaking their stores for the new system. The long-box provided consumers with something symbolic to be angry about, and the handful of international companies that dominate recorded music agreed on a symbolic response to quell their discontent. None of it was very important, but, with the exception of record retailers who had to reconfigure their stores and install new security systems, it made most of those involved feel good.

The long-box was one of the most psychologically misconceived packages ever. But the furor it triggered and its quick, effective result probably show why Americans are more engaged by solid-waste issues than they are by other, more serious pollution problems. The power they wield as consumers is more direct, and more likely to produce immediate results, than the power they have as citizens. Manufacturers are extremely sensitive to concerns about packages, which embody and express the product they contain. Packaging is, quite literally, a product's face and a company's good name. Moreover, a greener package offers a kind of marketing opportunity that, for example, cleaning up a polluting factory does not provide.

Thus, consumers and manufacturers have made a tacit agreement to be really concerned about packaging waste. The result has been a major change in the weight, size, and composition of packages during the 1990s. And everyone is reassured that it is possible to make a better world through shopping.

Afterword

While I was in the midst of writing this book, I spent some time with my mother as she recovered from difficult surgery.

Traumatized, she seemed to reorganize her life to minimize risk and purge disappointment, and she expressed this most emphatically in what she ate.

One day, her sister brought a pot of stew. "The color is funny," she said to me after my aunt had left. "You eat it if you like." I ate the stew, which was delicious, while my mother ate a bowl of Campbell's Home Cookin' soup.

A few days later, a neighbor brought a pot of excellent chicken soup. "I don't like the way she seasons things," my mother told me. For her lunch, she ate Nutri-Grain snack bars. That same day she said she would like pot roast for dinner. As I set off for the store to buy the meat, she said that she wanted me to use a ready-mix product, including seasoning and a bag in which it would be cooked. She clearly did not trust me to cook it in my own way.

I was irritated and concerned. Why should she trust Campbell and Kellogg's and Lawry's more than she trusted her sister, her neighbor, or her son? People wanted to be generous to her and to express their love, but she gave them no way to do it. She seemed to be retreating into a world of packages, in which the name-brand products not only replaced salesclerks but all of those who wanted to give her support.

You could see that she clung to the packaged products because she feared losing her independence. This was understandable, but it seemed to jeopardize her relationships with people who could help her survive.

The real problem with packaging, I concluded, is not its impact on landfills, but what it does to people's ability to connect with one another. Packages promise predictable, risk-free satisfaction, without all the unpredictability and irritation that accompany dealing with other people. And once people are isolated in this way, the society will become less resilient. It will wait for packaged solutions that can be accepted or rejected. People will be unable to be open to new ideas, or to one another. They won't be able to work together to deal with their problems.

My mother had replaced me with a mix! Family values seemed but a memory, a sense of community just a delusion.

My pique over having my pot roast recipe rejected continued for weeks, manifesting itself in random outbursts of social criticism. My mother wouldn't let me peel some carrots, and now the world was going to hell in a hermetically sealed, tamper-evident flexible pouch.

Only much later did I start to become honest with myself. The sense of independence my mother sought through her packaged products was not an illusion. I had been replaced, to be sure. But the frozen product in the microwavable tray was not as difficult to live with as I am.

For one thing, it wasn't restless because it had a book to finish. It didn't disagree with her. She didn't have to pay attention to its feelings, nor use energy she didn't have to keep it entertained. And the package was predictable in a way I was not. She welcomed my presence, up to a point, but she and I had long lived very separate lives, hundreds of miles apart, and neither of us was ready to be much closer.

I might feel that, in principle, it is better for families to be more intimate, more cohesive, and care more for one another. But I'm really not ready to live that way. Too much freedom, too many opportunities, have to be sacrificed. Families are, at once, nurturing and stifling. They are good to be in, and good to get away from.

My mother's devotion to packaged products was, at first, an insult. But ultimately it was a gift. It gave me permission to leave, while knowing that she could feed herself and be properly nourished. It let us both be independent. And who am I to say that her package-granted independence is an illusion and mine isn't?

My worries about the pernicious social effects of packaging and packaged thinking persist, at least in theory. In practice, though, I am pleased that she can have her life, and I can have mine. Despite my uneasiness, I can only be grateful to live in a world of packages.

Sources and further reading

This book has its origins in a catalog essay the Cooper-Hewitt Museum in New York asked me to write on the post-1940 work of Donald Deskey. Although the exhibition subsequently changed in focus, and the catalog was never published, working with Deskey's papers and those of Donald Deskey Associates at the Cooper-Hewitt's archive was a revelation. For the first time, I saw packaging as a major cultural, economic, and social activity seemingly hidden in plain sight.

This research also made me aware of the relatively small number of nontechnical works on packaging, and particularly its history. Alec Davis's *Package & Print* is the preeminent source on the physical and technological development of containers, organized by medium. Even though it's more than sixty-six years old, and written in a tone that is more promotional than analytic, Richard B. Franken and Carroll B. Larabee's *Packages That Sell* probably remains the fundamental text on the development of the marketing dimensions of packaging. Daniel J. Boorstin's *The Americans: The Democratic Experience* is probably the best general account of the rise of consumer culture, and one of the few that gives packaging its due. Its delightful account of Gail Borden is the basis for that given here.

Even before beginning my Cooper-Hewitt work, I had been much impressed by James R. Beniger's *The Control Revolution*, and I alluded to his discussion of the pioneer packager Henry Crowell in an earlier book. His is, for me, the most persuasive view of how manufacturing, marketing, consumption, and values have operated and evolved to shape the culture of which we are a part. This is an area that has produced an enormous amount of scholarship in recent years, most of it more frankly ideological than Beniger's somewhat technocratic approach. Warren I. Susman's work, especially the essay on character and personality cited below, is essential. The work of Stuart and Elizabeth Ewen and the works and anthologies of Richard Wightman Fox and T. J. Jackson Lears are important sources for accounts of the rise of consumer society. I found Lizabeth Cohen's article on chain grocery stores and the poor to be a particularly useful examination of a subject for which puffery and polemics are more common.

Still, industry-sponsored efforts, such as William Greer's *America the Bountiful,* with its first-person recollections by early packaged-goods marketers, are indispensable. Most manufacturers and marketers attach greater importance to packaging than historians generally do. Since the 1930s, the American Management Association has been publishing books and pamphlets on the packaging industry that form an important record of how containers, transportation, and retailing have evolved in that time.

I did much of the research at the Hagley Museum and Library, which is on the site of the former Du Pont gunpowder works in Delaware. Thus, I was able to use its excellent collection of trade catalogs and the archives of organizations, including the American Iron and Steel Institute. It also has enormous holdings on the Du Pont Company, which researched and promoted self-service retailing as a way to sell first cellophane and later Mylar and other packaging materials. I also made use of the N. W. Ayer archive at the Museum of American History at the Smithsonian, especially the advertising agency's pioneering work with Uneeda's In-Er Seal biscuit package.

Walter Stern's *Handbook of Package Design Research* provides an excellent overview of most of the important methods and ideas in that field. There are many catalogs of package designs, of which those published every few years by Graphis in Zurich are probably the most comprehensive and respected. The rapid changes in package design ensure that such books have a relatively short shelf life, and with a few exceptions readers are left to do their own analyses. Jeffrey Meikle's essay in David A. Hanks's *Donald Deskey* is a rare historical look at package design, though it probably understates the influence of such nondesigners as psychological consultant Louis Cheskin in determining the look of such iconic packages as Tide. Meikle's book *Twentieth Century Limited* is a good account of the rise of industrial design. Steven Heller and Seymour Chwast's *Graphic Style* provides a context for studying the appearance of packages, which tend stylistically to trail fashions in other media or follow other paths.

270

For environmental issues, the place to begin is William Rathje and Cullen Murphy's *Rubbish,* whose virtue is that it is based on the excavation of real landfills rather than on presumptions about what is being placed in them.

Because packaging is part of nearly every industry, it is covered to some extent by trade magazines featuring everything from supermarkets and pharmaceuticals to chemical manufacturers and railroads. For the package-converting industry itself, I have found *Packaging,* published by Cahners (Des Plaines, Illinois), to be a good guide to current developments. Since I completed the book, its former editor, Gregg Erickson, has launched a newsletter, *Shelf Presence* (Buffalo Grove, Illinois), which concentrates on the communicative dimensions of packaging that are the chief concern of this book.

In addition to Sinclair Lewis's *Main Street* and *Babbitt,* which are cited in the text, William Dean Howells's *The Rise of Silas Lapham* is essential reading, both for its descriptions of early packaging and marketing and for its depiction of the human cost of failing to turn a commodity into a product. Philip K. Dick's *Ubik* is a more contemporary view of the protean nature of packaged goods and the promise and terror they contain.

What follows is a fuller list of sources that were useful in writing this book.

Alexander, Judd H. *In Defense of Garbage.* Westport, Conn.: Praeger, 1993.

American Management Association. *Packaging, Packing & Shipping.* New York: American Management Association, 1936.

———. *Planning for Tomorrow's Packaging Realities.* New York: American Management Association, 1969.

Ayer, N. W., & Son. *Forty Years of Advertising.* Philadelphia: N. W. Ayer & Son, 1909.

Badler, Virginia R., Patrick E. McGovern, and Rudolph H. Michel. "Drink and Be Merry, Infrared Spectroscopy and Ancient Near Eastern Wine." In *MASCA Research Papers in Science and Archaeology* 7. Philadelphia: University Museum of the University of Pennsylvania, 1990.

Banham, Reyner. "Coffin-Nails in Handy Packs." In *Design by Choice.* London: Academy Editions, 1981.

Barber, Edwin A. *American Glassware.* Philadelphia: Patterson & White, 1900.

Bayley, Steven, Philippe Garner, and Deyan Sudjuc. *Twentieth-Century Style & Design.* New York: Van Nostrand Reinhold, 1986.

Bender, May. *Package Design and Social Change.* New York: American Management Association, 1975.

Beniger, James R. *The Control Revolution.* Cambridge: Harvard University Press, 1986.

Berndt, Heinrich. "The Energetics of the Bumblebee." In *The Insects,* by Thomas Eisner and E. O. Wilson. San Francisco: W. H. Freeman, 1977.

Birren, Faber. *Selling with Color.* New York: McGraw-Hill, 1945.

Black, W. H. *The Family Income.* New York: Delineator, 1907.

Boorstin, Daniel J. *The Americans: The Democratic Experience.* New York: Random House, 1973.

Buddenseig, Tilmann. *Industriekultur: Peter Behrens and the AEG.* Cambridge: MIT Press, 1984.

Burnett & Company. *The Law of Trademarks.* New York: Burnett & Co., 1868.

Cadman, S. Parkes. *A Prince in Commerce, a Master Builder.* Private ed. Eulogy to Frank W. Woolworth, Central Congregational Church. Brooklyn, 1919. In Hagley Museum and Library, Greenville, Del.

Calkins, Ernest Elmo. *The Business of Advertising.* New York: D. Appleton & Co., 1920.

————. *"Louder Please": The Autobiography of a Deaf Man.* Boston: Atlantic Monthly Press, 1924.

Campbell, Hannah. *Why Did They Name It.* New York: Fleet Publishing, 1964.

Can Manufacturers Institute. *The Metal Can/Its Past Present and Future.* Washington, D.C.: Can Manufacturers Institute, 1962.

Carson, Gerald. *The Old Country Store.* New York: Oxford University Press, 1954.

————. *One for a Man, Two for a Horse.* Garden City, N.Y.: Doubleday, 1961.

A Century of Trade Marks 1876–1976. London: Her Majesty's Stationery Office, 1976.

Chajet, Clive. *Image by Design.* Reading, Mass.: Addison-Wesley, 1991.

Chapman, Clowrey. *Trade-Marks.* New York: Harper & Brothers, 1930.

Charlton, D. E. A. *The Art of Packaging.* London: Studio, 1937.

Cheskin + Masten. *The Design Experience Model.* Pamphlet series. Redwood City, Calif.: Cheskin + Masten, 1992.

Cleary, Donald Powers. *Great American Brands.* New York: Fairchild Publications, 1981.

Cliff, Stafford. *The Best in Specialist Packaging Design.* Mies, Switzerland: Rotovision, 1992.

Cohen, Lizabeth. "The Class Experience of Mass Consumption." In *The Power of Culture,* by Richard Wightman Fox and T. J. Jackson Lears. Chicago: University of Chicago Press, 1993.

Crouwel, Wim, and Kurt Weiderman, eds. *Packaging: An International Survey.* New York: Praeger, 1968.

Davis, Alec. *Package & Print: The Development of Container and Label Design.* London: Faber & Faber, 1967.

Dennison, Richard A. "Stopping Packaging Waste at the Source."*Management Review,* June 1992, pp. 19–24.

deNoblet, Jocelyn, ed. *Industrial Design: Reflection of a Century.* Paris: Flammarion, 1993.

Deveneau, Willard F., ed. *History of Packaging during World War II: Plastic Packages and Materials.* Washington, D.C.: Packaging Institute, 1945.

Dichter, Ernest. *Packaging: The Sixth Sense.* Boston: Cahners Books, 1975.

Dick, Philip K. *Ubik.* 1969. Reprint, New York: Vintage Books, 1991.

Dolan, J. R. *The Yankee Peddlers of Early America.* New York: Clarkson N. Potter, 1964.

Driscoll, David B. "Gunpowder Containers in the Nineteenth Century: The Drive for Uniformity." Greenville, Del.: Hagley Museum and Library, 1984.

Drummond, J. C., and W. R. Lewis. *The Examination of Some Tinned Foods of Historic Interest.* Greenford, Middlesex: Publications of the International Tin Research and Development Council, 1938.

Dubois, J. Harry. *Plastics History USA.* Boston: Cahners Books, 1972.

Du Pont Cellophane Co. *Color, Pattern, Texture . . . Your Greatest Selling Aids.* Wilmington, Del.: Du Pont Cellophane Co., 1933.

Editors of Consumer Reports. *I'll Buy That!* Mount Vernon, N.Y.: Consumers Union, 1986.

Ewen, Stuart.*Captains of Consciousness.* New York: McGraw-Hill, 1976.

Ewen, Stuart and Elizabeth. *Channels of Desire: Mass Images and the Shaping of American Consciousness.* Minneapolis: University of Minnesota Press, 1992.

Favre, Jean-Paul. *Color Sells Your Package.* Zurich: ABC Edition, 1969.

Forty, Adrian. *Objects of Desire.* New York: Pantheon, 1986.

Franken, Richard B., and Carroll B. Larabee. *Packages That Sell.* New York: Harper & Brothers, 1928.

Gerstman + Meyers. *Consumer Solid Waste: Attitude and Behavior Study V.* New York: Gerstman + Meyers, 1993.

Glass Container Manufacturers Institute. *The Story of Glass Containers.* Washington, D.C.: Glass Container Manufacturers Institute, 1960.

Great Atlantic & Pacific Tea Co. *You and Your Company*. Handbook for store managers. New York: Great Atlantic & Pacific Tea Co., 1944.

Greer, William. *America the Bountiful: How the Supermarket Came to Main Street*. Washington, D.C.: Food Marketing Institute, 1986.

Guss, Leonard M. *Packaging Is Marketing*. New York: American Management Association, 1966.

Harris, Neil. *Cultural Excursions*. Chicago: University of Chicago Press, 1990.

Hayden, Dolores. *The Grand Domestic Revolution*. Cambridge: MIT Press, 1981.

Heller, Steven, and Seymour Chwast. *Graphic Style*. New York: Harry N. Abrams, 1988.

Herdeg, Walter. *Packaging, Packungen, Emballages*. Zurich: Graphis Press, 1959 and subsequent editions.

Hirshorn, Paul, and Steven Izenour. *White Towers*. Cambridge: MIT Press, 1979.

Holland, D. K. *Great Packaging*. Rockport, Mass.: Rockport Publishers, 1992.

Howells, William Dean. *The Rise of Silas Lapham*. 1885. Reprint, New York: Library of America, 1982.

Jacobson, Egbert, ed. *Seven Designers Look at Trademark Design*. Chicago: Paul Theobald, 1952.

Jay, Leslie. "The Dastardly Dozen, 12 Recycling Myths Debunked." *Management Review*, June 1992, pp. 5–12.

Kamekura, Yusaku. *Trademarks of the World*. New York: George Wittborn, 1956.

Katz, Sylvia. *Plastics*. New York: Harry N. Abrams, 1984.

Kaufman, William T. *Perfume*. New York: Dutton, 1974.

Ketchum, William C. *Boxes*. New York: Cooper-Hewitt Museum, 1982.

Langdon, Philip. *Orange Roofs, Golden Arches*. New York: Alfred A. Knopf, 1985.

Lears, T. J. Jackson. "From Salvation to Self-Realization: Advertising and the Therapeutic Roots of American Consumer Culture." In *The Culture of Consumption*, ed. Richard Wightman Fox and T. J. Jackson Lears. New York: Pantheon, 1983.

————. *No Place of Grace*. New York: Pantheon, 1981.

Levitt, Theodore. "Production-line Approach to Service" and "The Industrialization of Service." In *Levitt on Marketing*. Cambridge: Harvard Business School Press, 1991.

Lewis, Sinclair. *Babbitt*. New York: Harcourt Brace, 1922.

————. *Main Street*. New York: Harcourt, Brace and Howe, 1920.

Liebs, Chester H. *Main Street to Miracle Mile.* Boston: New York Graphic Society, 1985.

Lief, Alfred. *It Floats: The Story of Procter and Gamble.* New York: Rinehart & Co., 1958.

———. *The Mennen Story.* New York: McGraw-Hill, 1954.

Little, Arthur D., Inc. *The Role of Packaging in the U.S. Economy.* New York: American Foundation of Management Research, 1966.

Loden, D. John. *Megabrands: How to Build Them, How to Beat Them.* Homewood, Ill.: Business One Irwin, 1992.

Loewy, Raymond. *Industrial Design.* Woodstock, N.Y.: Overlook Press, 1979.

Luxenberg, Stan. *Roadside Empires.* New York: Viking, 1985.

Lynch, Kevin. *Wasting Away.* San Francisco: Sierra Club Books, 1990.

McGrath, Molly Wade. *Top Sellers, USA.* New York: Morrow, 1983.

Marquette, Arthur F. *Brands, Trade Marks and Good Will.* New York: McGraw-Hill, 1967.

Meikle, Jeffrey L. *Twentieth Century Limited: Industrial Design in America 1925–1939.* Philadelphia: Temple University Press, 1979.

———. "Donald Deskey Associates: The Evolution of an Industrial Design Firm." In *Donald Deskey: Decorative Arts and Interiors,* by David A. Hanks. New York: Dutton, 1987.

Myers, William. *The Image Makers.* New York: Times Books, 1984.

Neaubauer, Robert G. *Packaging: The Contemporary Media.* New York: Van Nostrand, 1973.

Nichols, John P. *The Chain Store Tells Its Story.* New York: The Institute of Distribution, 1940.

———. *Skyline Queen and the Merchant Prince: The Woolworth Story.* New York: Trident Press, 1973.

Opie, Robert. *Packaging Source Book: A Visual Guide to a Century of Packaging Design.* Secaucus, N.J.: Chartwell Books, 1989.

Oswin, C. R. *Plastic Films and Packaging.* New York: John Wiley, 1975.

Packaging, Report of a Specialist Team Which Visited the United States of America in 1950. London: Anglo-American Council of Productivity, 1950.

Packard, Vance. *The Hidden Persuaders.* Rev. ed. New York: Pocket Books, 1977.

Patten, Simon N. *The Consumption of Wealth.* Philadelphia: Publications of the University of Pennsylvania, 1889.

Price, Peter W. *Insect Ecology.* New York: John Wiley, 1975.

Pulos, Arthur J. *The American Design Adventure.* Cambridge: MIT Press, 1988.

Rathje, William, and Cullen Murphy. *Rubbish! The Archaeology of Garbage.* New York: HarperCollins, 1992.

The Rauch Guide to the U.S. Packaging Industry. Annual. Bridgewater, N.J.: Rauch Associates.

Roth, Laszlo, and George Wybenga. *The Packaging Designer's Book of Patterns.* New York: Van Nostrand Reinhold, 1991.

Sacharow, Stanley. *The Package as a Marketing Tool.* Radnor, Penn.: Chilton Book, 1982.

————. *Packaging Design.* New York: PBC International, 1983.

Saitoh, Hideo. *Can Design.* Tokyo: Bijutsu Shuppan-Sha, 1991.

————. *Carton Design.* Tokyo: Bijutsu Shuppan-Sha, 1992.

Schisgall, Oscar. *Eyes on Tomorrow: The Evolution of Procter and Gamble.* Chicago: J. G. Ferguson Publishing, 1981.

Selame, Elinor and Joe. *Packaging Power: Corporate Identity and Product Recognition.* New York: American Management Association, 1982.

Shaw, Robert B. *History of the Comstock Patent Medicine Business and Dr. Morse's Indian Root Pills.* Washington, D.C.: Smithsonian Institution Press, 1972.

Shelton, Vaughan. *Steel Came to Dinner.* Washington, D.C.: American Iron and Steel Institute, 1957.

Smith, Clayton Lindsay. *The History of Trademarks.* New York: Printed by Thomas H. Stuart, 1923.

Smith, Terry. *Making the Modern: Industry, Art, and Design in America.* Chicago: University of Chicago Press, 1993.

Stern, Walter, ed. *Handbook of Package Design Research.* New York: Wiley-Interscience, 1981.

Susman, Warren I. *Culture as History.* New York: Pantheon, 1984 (particularly the essay " 'Personality' and the Making of Twentieth Century Culture").

Sutnar, Ladislav. *Package Design: The Force of Visual Selling.* New York: Arts Inc., 1953.

Swank, James M. *Our Tinplate Industry Built Up by the McKinley Tariff.* American Iron and Steel Institute Tariff Tract No. 2. Washington, D.C.: American Iron and Steel Institute, 1896.

Tedlow, Richard S. *New and Improved: The Story of Mass Marketing in America.* New York: Basic Books, 1990.

Thomson, Henry C. *Trademarks.* Boston: Henry C. Thompson, 1914.

United States Environmental Protection Agency. *Life Cycle Design Guidance Manual.* Washington, D.C.: EPA, 1993.

The Value of Advertised Trademarks. Ben B. Hampton Co., 1903. Originally published in *Printer's Ink* (1900–1902).

Walsh, William I. *The Rise and Decline of the Great Atlantic & Pacific Tea Company.* Secaucus, N.J.: Lyle Stuart, 1986.

Wilson, Edward O. *On Human Nature.* Cambridge: Harvard University Press, 1978.

Wilson, Richard Guy, Diane Pilgrim, and Richard Tashjian. *The Machine Age in America 1918–1941.* New York: Harry N. Abrams, 1986.

Wilson, Terry P. *The Cart That Changed the World: The Career of Sylvan N. Goldman.* Norman: University of Oklahoma Press, 1978.

Woolworth, F. W., Co. *Fortieth Anniversary Souvenir.* New York: F. W. Woolworth Co., 1919.

———. *Home Shopping Guide: Nothing over 10c.* New York: F. W. Woolworth Co., 1929.

Yao, Takeo. *Package Design in New York.* Tokyo: Nippon Shuppan Hanbai, 1985.

———. *Package Design in Tokyo.* Tokyo: Nippon Shuppan Hanbai, 1987.

Young, James Harvey. *The Toadstool Millionaires.* Princeton, N.J.: Princeton University Press, 1961.

Acknowledgments

*T*hanks, first of all, to those who helped inspire the idea:

The woman, whose name I didn't write down, who telephoned me in 1986, after reading my book *Populuxe,* and said I should write next about packaging. I ignored her.

My colleague at the *Philadelphia Inquirer,* Lesley Valdes, who looked at the jacket for my second book, *Facing Tomorrow,* and said, with despair-inducing accuracy, "Why does your book look like a cigarette pack?" It wasn't Marlboro, alas, or even Viceroy.

Most important, Gail Davidson, of the Cooper-Hewitt Museum, who really got me going by commissioning me to write an essay.

And thanks to those in the fields of package design and research who shared their insights, among them Paul Spiker, Elinor and Joseph Selame, Richard Gerstman, Ronald Peterson, John Lister, Anita Hersh, Davis Masten, Christopher Ireland, Gregory Hing, Alan Cutler, the late Takeo Yao, Katsura Iwamatsu, Motoo Nakamishi, Del Macaulay, Stan Gross, Robert Colonna, and Ric Hirst. In a class by himself is Irv Koons, a pioneering package designer, who made his recollections, his class notes, and photographs of his package collection available for this book.

Rosalind Berman allowed me to examine and photograph her excellent collection of old medicine and grocery packages. Michael Mally photographed many packages, and packaged the author. Kinya Murayama very generously opened Japan for me. Patrick McGovern provided an archaeological perspective on packaging. Mark Lankin and his colleagues at the Hagley Museum and Library were tremendously helpful in finding and suggesting materials to examine. Glenn Gunderson, of Dechert Price & Rhoads, provided a valuable legal briefing and reading suggestions.

Stephan Salisbury, another *Inquirer* colleague, was present for the beginning and end of this project, and was its first reader. His own researches on debt and consumption led me in some interesting directions. Barney Karpfinger, my agent, was an enthusiastic advocate who helped prod me into action.

Index